CHINESE REMAINDER THEOREM

Applications in Computing, Coding, Cryptography

CHINESE REMAINDER THEOREM

Applications in Computing, Coding, Cryptography

C. Ding
Turku Centre for Computer Science,
Turku, Finland

D. Pei
Chinese Academy of Science,
Beijing, China

A. Salomaa
Department of Mathematics,
University of Turku, Turku, Finland

World Scientific
Singapore • New Jersey • London • Hong Kong

Published by

World Scientific Publishing Co. Pte. Ltd.

5 Toh Tuck Link, Singapore 596224

USA office: 27 Warren Street, Suite 401-402, Hackensack, NJ 07601

UK office: 57 Shelton Street, Covent Garden, London WC2H 9HE

British Library Cataloguing-in-Publication Data
A catalogue record for this book is available from the British Library.

CHINESE REMAINDER THEOREM
Applications in Computing, Coding, Cryptography

ISBN-13 978-981-02-2827-9
ISBN-10 981-02-2827-9

Preface

Chinese Remainder Theorem, CRT, is one of the jewels of mathematics. It is a perfect combination of beauty and utility or, in the words of Horace, *omne tulit punctum qui miscuit utile dulci.* Known already for ages, CRT continues to present itself in new contexts and open vistas for new types of applications. So far, its usefulness has been obvious within the realm of "three C's". Computing was its original field of applications, and continues to be important as regards various aspects of algorithmics and modular computations. Theory of codes and cryptography are two more recent fields of application.

This book tells about CRT, its background and philosophy, history, generalizations and, most importantly, its applications. The book is self-contained. This means that no factual knowledge is assumed on the part of the reader. We even provide brief tutorials on relevant subjects, algebra and information theory. However, some mathematical maturity is surely a prerequisite, as our presentation is at an advanced undergraduate or beginning graduate level. We have tried to make the exposition innovative, many of the individual results being new.

A special course about CRT can be based on the book. The individual chapters are largely independent and, consequently, the book can be used as supplementary material for courses in algorithmics, coding theory, cryptography or the theory of computing. Of course, the book is also a reference for matters dealing with CRT.

Acknowledgements: We would like to thank the Academy of Finland, Chinese Academy of Science, and the Turku Centre for Computer Science (officially listed as one of the top twelve research centers in Finland) for providing us with excellent working conditions and financial support for our research. Special thanks are due to Elisa Mikkola for her assistance in several aspects of the project. Also we want to thank World Scientific, and especially Mr. Richard Lim, for good cooperation in this book project. We would also like to thank all members of our families for their support.

August 1996,

Cunsheng Ding   Dingyi Pei   Arto Salomaa

Contents

Chapter 1

Introduction and Philosophy

The story about the Chinese Remainder Theorem, CRT, can be told in many ways. In a highly abstracted mathematical exposition, one would start with a Dedekind ring and the decomposition of the principal ideal into a product of prime ideals and then proceed to the so-called Theorem on the Independence of Exponents which can be viewed as a possibility for an abstract formulation of CRT. Such an approach would forget the original landscape of CRT: integers and remainders under division. The opposite extreme of telling the story would consist of various numerical examples in the original landscape, with little general theory, or none whatsoever.

Our approach in this book lies between the two extreme approaches mentioned. Indeed it can be labeled as applications-oriented. During its long history, CRT has appeared in many disguises, never failing to find new aspects of application. An aspect inherent in the very core of CRT is *computing:* algorithms for taking calculations via a detour where much smaller numbers can be used. Such a detour can be partitioned in a fashion where redundancy is added to the data, or data can be recovered only with the cooperation of sufficiently many parties sharing them partially. Then aspects of (error-correcting) *codes* and *cryptography* become relevant. This book tells the story of CRT in the landscape of three C's: computing, codes and cryptography. We feel that our approach in some sense reflects the change visible in mathematical research in general. There is greater interest in constructive algorithmic results, also in their efficiency, and less interest in only existential studies with little or no computational significance.

1.1   A Historical Overview

The Chinese Remainder Theorem, CRT, appeared in the mathematical classic of Sun Zi, a mathematician in ancient China. The book is known by the name

Sun Zi Suanjing, Sun's Arithmetical Manual. The exact date is unknown, but it is reasonable to take it to be during the first century A. D. Although we do not plan to dwell on the history of mathematics, it is useful to try to get some time-perspective.

Every human culture exhibits mathematics, at least in some primitive forms. "Western" mathematics, as a systematic pursuit, originated in Egypt and Mesopotamia, achieved an early culmination in Greece and spread to the Graeco-Roman world. After the fall of Rome, there was a stillstand in mathematical creativity in Europe, lasting half a millennium. On the other hand, the Islamic branch was born, to become combined with the European branches. Thus, we arrive at the following rough timetable for the periods of development of western mathematics:

Egyptian	3000 B.C.—1500 B.C.
Babylonian	1700 B.C.—300 B.C.
Greek	600 B.C.—200 B.C.
Graeco-Roman	A.D. 100—A.D. 500
Islamic	A.D. 750—A.D. 1450
Medieval-Renaissance	A.D. 1100—A.D. 1600
Modern	A.D. 1600—

Early *oriental* and western mathematics were quite isolated from one another. Details of possible interactions are not clear and are still a subject of further investigations. It can be seen from the above timetable that the book of CRT, *Sun Zi Suanjing*, falls timewise to the beginning of the Graeco-Roman period in western mathematics. Let us now briefly look into the history of early Chinese mathematics, before and during the time of *Sun Zi Suanjing*, to get a comparison of the above western timetable.

The oldest Chinese mathematical classic is *Chou Pei Suanjing*, a record of mathematics for astronomical calculations from about 1000 B.C. The Pythagorean Theorem was already used in the astronomical calculations of this book.

The most influential of all ancient Chinese mathematical books was *Jiuzhang Suanshu*, Nine Chapters on the Mathematical Art. It was composed about A.D. 50-100, somewhat earlier than *Sun Zi Suanjing*. It includes 246 problems and solutions coming from practice. The calculation of square and cubic roots can be found in some of the solutions. A systematic method for solving some systems of linear equations, involving also negative numbers, is presented in the book. The last chapter includes results on rectangular triangles, some of which were rediscovered later in India and Europe. In this book the approximation for π equals 3. Later a Chinese geometer Liu Hui, an important commentator on this book, improved the value of π to 3.14 by considering a regular polygon of

96 sides, and further to 3.14159 by considering a polygon of 3072 sides. The Chinese mathematician Tsu Chung-Chih (430-501) knew the approximation $\pi \approx 22/7$ and called it "inexact", and presented also the more accurate value $\pi \approx 355/113 \approx 3.1415929$. Basically, the main topics in *Sun Zi Suanjing* are the same as those in the Nine Chapters, *Jiuzhang Suanshu* , except that one topic appears for the first time in Sun's Manual: The Chinese Remainder Theorem.

Finally, it should be emphasized that the relative isolation of early Chinese and western mathematics began to decrease later on. Nowadays mathematicians form a worldwide community whose unification will become even stronger during the times of the Internet.

1.2 Pars pro toto

The Chinese Remainder Theorem can be viewed as a manifestation of the general principle "pars pro toto"—a part goes for the whole thing. Aspects of this principle can be found in most different environments. We now try to illustrate the principle *pars pro toto* by some examples, converging towards our ultimate goal, the Chinese Remainder Theorem. The crucial question in this connection will be: can the whole thing be replaced or represented by its part, or to what extent or in what sense can it be so replaced or represented?

An area where the principle *pars pro toto* appears in a fascinating form in *genetics*. The whole individual is in a very definite sense present in a single cell or in some DNA strips. Under favorable circumstances, these DNA strips can still much later be recovered in fossils remaining after the individual.

Another example of the principle *pars pro toto*, lying in quite a different direction, is decision-making through *parliamentary democracy*. Instead of direct decision-making, a population elects representatives, members of a parliament, who are going to make the decisions. Thus, the whole population is represented or replaced by a part, the parliament, for the process of decision-making. This book is not an appropriate place to discuss the virtues or disadvantages of such a representation, for instance, as regards the various minorities in the population. Instead, we proceed to an example coming closer to our actual topic of CRT: how well can *information* or *data* be represented by its suitably chosen parts?

There are many instances of situations, where a thing is or should be found out by some of its properties. Thus, in such a case the properties in question contain enough information to identify the whole thing—the whole thing is represented or can be replaced by those properties. Having a property can be understood as belonging to the class of objects with this property. As regards each specific class S and object X, the matter is settled by a "yes" or "no" answer to the question "Is X in S?" In a popular game of questions and answers, one has to find the identity of an object X by such questions whose number should not exceed a pregiven bound, say twenty.

Let us take one step forward and become more specific. By a suitable encoding, all information can be represented as numbers. So let us assume that our object X, the piece of information we are interested in, is a positive integer. We are allowed to ask questions of the form

(n) Is X greater than n?

In this way we get a sequence of answers of the form (n, Z), where Z =yes or Z =no. (For instance, (1000, yes) says that X is greater than 1000.) It is obvious that any X is determined by a sequence of answers

$$(n_1, Z_1)(n_2, Z_2) \cdots (n_t, Z_t).$$

A trivial way to do this is to consider the sequence

$$(n, \text{no})(n - 1, \text{yes}),$$

which identifies X as n. However, if X is completely unknown, we cannot possibly expect to be so lucky that we would immediately guess the questions (n) and ($n - 1$)!

A general strategy would be the following. First one has to aim at a "no" answer. After an answer (n, no), roughly $\log_2 n$ questions are needed to determine X. This matter will be discussed further from an information-theoretic point of view in Section 4.1. For instance, the sequence of answers

(128, no)(64, yes)(96, no)(80, yes)(88, yes)(92, yes)(94, yes)(95, yes)

determines the value $X = 96$, the number of questions after the first "no"-answer being $\log_2 128 = 7$. For convenience we have only used powers of 2. The first questions, before the first "no"-answer, can be only guesses.

In the above example, the properties we used for characterizing the number X were formulated by question (n) and, thus, were very simple indeed. They are simple also in the characterization due to the Chinese Remainder Theorem, CRT. An unknown number X is characterized by its remainders under divisions by different integers n. Thus, the questions asked will be of the form

(n) What is the smallest nonnegative remainder of X modulo n?

Here it is more convenient to formulate the questions in this way but, if so preferred, each question (n) can obviously be replaced by n questions with only "yes" or "no" answers.

Does such a *pars pro toto* representation, with suitably chosen moduli, always determine a unique X, that is, can a number X always be characterized in this way by remainders? Obviously, this is not possible because, no matter how the moduli $n_1, \cdots, n_t$ are chosen, each of the infinitely many numbers

$$X + in_1 \cdots n_t, \quad \text{for } i = 0, 1, 2, \cdots,$$

possesses the same smallest nonnegative remainders with respect to each of the moduli. As we will see in the exact formulation of the Chinese Remainder Theorem, this method of representation by remainders determines a unique X only among numbers having a well-specified size.

1.3 Chinese Remainder Theorem: A First Formulation

A magician wants to impress the audience with the following "mind-reading" trick. A randomly chosen helper is asked to think of a number less than 60. Then he/she is asked to tell the remainders when the number is divided by 3, 4, and 5, in succession. Upon hearing the remainders, the magician tells the number. For instance, the number will be 37 in case of the remainders 1, 1 and 2, obtained in divisions by 3, 4, and 5, respectively.

This illustration is taken from an old guide-book for magicians. The book tells the magician to divide the number $40a + 45b + 36c$, where a, b and c are the three remainders, by 60 and announce the remainder as the result of the mind-reading! In the example case the division of 157 by 60 leaves indeed the remainder 37. Although this example is correctly treated, the book is not very successful in its generalization to arbitrary moduli.

Let us now do it properly and prove the Chinese Remainder Theorem in its basic version. We assume here that the reader is familiar with the fact that the greatest common divisor of two integers a and b, denoted by $\gcd(a, b)$, can be represented as the sum

$$\gcd(a, b) = ua + vb,$$

for some integers u and v, following the so called Euclidean algorithm. We also want to point out that the discussion of the basic version of CRT will be resumed in Chapter 2 from a slightly different angle. In this way we hope to give the reader a solid background before moving into the more advanced chapters.

We say that the integers $m_1, m_2, \cdots, m_t$ are *relatively prime in pairs* if $\gcd(m_i, m_j) = 1$ for any distinct i and j where $1 \leq i, j \leq t$.

Chinese Remainder Theorem Let $a_1, a_2, \cdots, a_t$ be any t integers and $m_1, m_2, \cdots, m_t$ be relatively prime in pairs. Then there is a number x with the property

$$x \equiv a_i \quad (\text{mod } m_i), \quad i = 1, 2, \cdots, t. \tag{1.1}$$

Moreover, x is unique in the following sense. Let M be the product $m_1 m_2 \cdots m_t$ and let y satisfy the system of congruences (1.1). Then $y \equiv x \pmod{M}$.

Proof: We apply induction on t. For $t = 1$, it suffices to take $x = a_1$. The uniqueness as asserted is also obvious.

Assumingly that the result holds for $t = k$, we demonstrate its validity for $t = k + 1$. Consider the system of congruences

$$x \equiv a_1 \quad (\text{mod } m_1),$$
$$\cdots\cdots\cdots$$
$$x \equiv a_k \quad (\text{mod } m_k),$$
$$x \equiv a_{k+1} \quad (\text{mod } m_{k+1}),$$

where the moduli are relatively prime in pairs. By the inductive hypothesis, there is a number x' satisfying the first k of these congruences.

Observe that the product $m_1 \cdots m_k$ and the integer m_{k+1} are relatively prime, that is, their greatest common divisor equals 1. Otherwise, they would have a common prime factor p. Since p divides the product $m_1 \cdots m_k$, it must divide one of the factors, say m_j, $1 \leq j \leq k$. But then

$$\gcd(m_j, m_{k+1}) \geq p > 1,$$

which contradicts the assumption of the moduli being relatively prime in pairs.

Since $m_1 \cdots m_t$ and m_{k+1} are relatively prime, there are integers u and v (possibly negative) such that

$$um_1 \cdots m_k + vm_{k+1} = 1.$$

Multiplying both sides by $(a_{k+1} - x')$, we obtain

$$u(a_{k+1} - x')m_1 \cdots m_k + v(a_{k+1} - x')m_{k+1} = a_{k+1} - x'$$

and, further,

$$x' + u''m_1 \cdots m_k = a_{k+1} + v''m_{k+1},$$

where we have abbreviated

$$u'' = u(a_{k+1} - x') \text{ and } v'' = -v(a_{k+1} - x').$$

Denoting $x'' = x' + u''m_1 \cdots m_k$, this tells us that

$$x'' \equiv a_{k+1} \pmod{m_{k+1}}.$$

On the other hand,

$$x'' \equiv x' \equiv a_i \pmod{m_i}, \quad i = 1, \cdots, k,$$

where the first congruence follows by the definition of x'', and the second congruence is due to the choice of x'. Thus, x'' satisfies all of our original $k + 1$ congruences, and we have completed the induction step.

For the assertion about uniqueness, consider two numbers x_1 and x_2 satisfying (1.1). Thus, also

$$x_1 \equiv x_2 \pmod{m_i}, \quad i = 1, 2, \cdots, t,$$

from which the relation $x_1 \equiv x_2 \pmod{M}$ immediately follows, by the definition of a congruence and by our assumption about these m_i being relatively prime in pairs. □

Although the computational aspect is not emphasized in the above proof, the proof still contains the ingredients of both the iterative and direct Chinese Remainder Algorithm, as will become clear from the discussion in Chapter 2.

As a first indication of the power of the Chinese Remainder Theorem, we prove a corollary showing how any finite sequence of integers can be represented in terms of two integers. The corollary is important in various considerations dealing with logic and mathematics, but it will not be needed as such in this book. We will use here and later on in this book the short notation $(a \bmod n)$, with or without parentheses, for the smallest nonnegative remainder of a modulo n.

Corollary Let a_i, $0 \le i \le t$, be a finite sequence of nonnegative integers. Then there are integers u and v such that

$$(u \bmod (1 + (i + 1)v)) = a_i, \quad \text{for every } i = 0, 1, \cdots, t.$$

Proof: Let a be the largest among the integers a_i, $0 \le i \le t$, and define $v = 2a \cdot t!$ and $m_i = 1 + v(i + 1)$, $0 \le i \le t$. We claim that the integers m_i, $0 \le i \le t$, are relatively prime in pairs. Assume the contrary: a prime p divides both m_i and m_j, for some $i > j$. Then p divides also the difference

$$(i + 1)m_j - (j + 1)m_i = i - j \le t.$$

Since $p \le t$ divides m_i and v is divisible by all integers $\le t$, we obtain the contradiction $p = 1$. This proves our claim.

Thus, the integers m_i qualify as moduli for the Chinese Remainder Theorem. Hence, there is a number u such that

$$u \equiv a_i \pmod{m_i}, \quad i = 0, 1, \cdots, t.$$

Hence, $(u \bmod m_i) = (a_i \bmod m_i)$, for all $0 \le i \le t$. But because $a_i < v < m_i$, we conclude that

$$(a_i \bmod m_i) = a_i, \quad \text{for } 0 \le i \le t.$$

Hence,

$$(u \bmod m_i) = a_i, \quad \text{for } 0 \le i \le t,$$

as asserted in our corollary. $\qquad\qquad\qquad\qquad\qquad\qquad\qquad\qquad\qquad\qquad\square$

1.4   CRT in the Hands of Old Mathematicians

The view that mathematics is a cumulative science in the sense that earlier results are needed in building up later theories is only partially true. Some entire theories become obsolete and unpopular, and pass into oblivion. The advent of computers has very much changed the map of mathematics in making the *discrete modeling* of the world a very feasible and most applications-oriented approach. As a result, for instance, *graph theory* has gained tremendously in prestige and has become a huge science with numerous big branches, from an earlier slum area of topology.

With an estimated one million (more or less) new mathematical theorems being established every year, one cannot expect that even most of them will be useful in later developments. Very often an area of mathematics becomes saturated, after which research stubbornly pursued in this area is bound to remain on small side tracks. When remnants of such research are later discovered in some connection, this does not usually happen with enthusiasm. Doors of the mathematical past being rusted does not mean that there lies a treasure inside.

However, there are many marvelous exceptions. Some research areas and individual results seem to thrive in most diverse mathematical cultures. Such areas are repeatedly used as a basis of new fields of research, and such results customarily pop up in various disguises during the course of history. Basic number theory is certainly such a research area, and CRT such an individual result.

We have already spoken about the origins of CRT. The Chinese background will be discussed in Chapter 2. We now give glimpses about how CRT is treated by two old mathematicians, Fibonacci and Euler. Fibonacci, Filius Bonacci, the son of Bonaccus, also known as Leonardo Pisano, wrote in his *Liber Abbaci* from 1202 roughly as follows.

"Let a contrived number be divided by 3, also by 5, also by 7; and ask each time what remains from each division. For each unity that remains from the division by 3, retain 70; for each unity that remains from the division by 5, retain 21; and for each unity that remains from the division by 7, retain 15. And as much as the number surpasses 105, subtract from it 105; and what remains to you is the contrived number. Example: suppose from the division by 3 the remainder is 2; for this you retain twice 70, or 140; from which you subtract 105, and 35 remains. From the division by 5, the remainder is 3; for which you retain three times 21, or 63, which you add to the above 35; you get 98. From the division by 7, the remainder is 4, for which you retain four times 15, or 60; which you add to the above 98, and you get 158, from which you subtract 105, and the remainder is 53, which is the contrived number. From this rule comes a pleasant game, namely if someone has learned this rule with you; if somebody else should say some number privately to him, then your companion, not interrogated, should silently divide the number for himself by 3, by 5, and by 7 according to the above-mentioned rule; the remainders from each of these divisions he says to you in order; and in this way you can know the number said to him in private."

Sun's original approach will be described in Chapter 2. In general mathematical terms, Fibonacci's presentation is very similar. Both of them have their presentation based on a specific numerical example, however, an implicit feeling is conveyed about the method being general. Neither one worries about the uniqueness of the solution. Observe the very pleasant leisurely writing style of Fibonacci. His approach is also very algorithmic and directly implementable. Rather surprising is the cryptographic touch in his description of the "pleasant game". Remember that cryptography is one of the C's in the landscape of this book.

From the point of view of mathematical terms and notation, the approach becomes very different in the writing of Leonhard Euler. The integers in the specific example are replaced by variables, and modern algebraic notation is followed. There is also no doubt that the method is intended to be general: instead of five numbers, there can be arbitrarily many. The text by Euler is from the publication of St. Petersburg Academy of Sciences in 1734:

"A number is to be found that, when divided by a, b, c, d, e, which numbers I suppose to be relatively prime, leaves respectively the remainders p, q, r, s, t. For this problem the following numbers satisfy:

$$Ap + Bq + Cr + Ds + Et + m \times abcde$$

in which A is the number that divided by $bcde$ has no remainder, by a, however, has the remainder 1; B is the number that divided by $acde$ has no remainder, by b, however, has the remainder 1, ..., which numbers can consequently be found by the rule given for two divisors."

Here the last remark refers to a result earlier in the text, needed in the computing of the numbers A, B, C, D, E.

We conclude this section with the following formulation of the Chinese Remainder Theorem in a ring-theoretic setting:

Let A be a commutative ring with identity and let $\{m_i | 1 \leq i \leq t\}$ be a finite collection of ideals in A such that $m_i + m_j = A$ for all $i \neq j$. Then, for any set of elements $\{x_i \in A | 1 \leq i \leq t\}$, there exists an $x \in A$ such that

$$x \equiv x_i \pmod{m_i}, \ 1 \leq i \leq t.$$

The reader should be able to figure out, eventually after consulting the Tutorial in Algebra given in Appendix B, how the basic numerical version of the Chinese Remainder Theorem results from the ring-theoretic formulation. Instead of explaining this matter in detail, we conclude this section with a story about Oscar Zariski, a foremost figure in the field of algebraic geometry, told in [28].

Oscar Zariski gave a course in projective geometry on a highly abstract level. One of the students, afterwards a well-known mathematician, felt the need of some clarification and examples. "What would you get", he asked the professor, "if you specialized the field F to the complex numbers?" Zariski answered: "Yes, just take F as the complex numbers."

1.5   CRT in Applications: the Three C's

We have already pointed out that we will not pursue in this book the abstract mathematical formulation when telling about CRT. Indeed, at a sufficiently high level of abstraction, CRT becomes an axiom for the structure you are investigating—and so there is very little you can say about it, but you have to proceed to other matters instead. This means that, instead of the title "CRT", the book should have a title such as "CRT-Based Structures".

But there is much to say about the *use* of CRT in various contexts. This is what our book is about. Specifically, we will tell about the applications of CRT in the landscape of three C's: computing, codes, cryptography.

We want to emphasize that the list of applications of CRT considered in this book is by no means exhaustive. Indeed, applications of CRT occur in almost every area of mathematics. We just mention the following modern applications to the theory of finite automata by Yu, Zhuang and Salomaa [105]. Let m and n be relatively prime. By a CRT-based argument concerning the largest integer not representable in the form $cm + dn$, for some positive or nonnegative c, d (three possible cases), it is shown in [105] that the representation of the catenation $L_1 L_2$ in a finite deterministic automaton may require mn states, if the representation of L_1 (resp. L_2) requires m (resp. n) states. Moreover, the number of states which is necessary and sufficient in the worst case for a finite deterministic automaton to accept the star of an n-state language, $n > 1$, over a one-letter alphabet is $(n-1)^2 + 1$. We neither explain nor use these notions in this book, so this illustration is meant to a reader knowledgeable in automata theory.

The Chinese Remainder Problem arose originally from the computation of calendars, as will be seen below in Chapter 2, where also many other ancient problems, related to modular computations, are discussed. The C of *computing*, in the landscape of three C's we are describing, is undoubtedly the oldest and most diversified. It will be the subject matter of Chapters 3-5 below. The Chinese Remainder Algorithm, CRA, is basically a divide-and-conquer technique. The originally given problem is divided into subproblems. The latter can be solved independently of each other, so parallel processing is possible. Finally, the results for the subproblems are combined by CRT to get a solution of the original problem. This procedure is depicted in Fig. 3.1 in Section 3.1 below.

We will not describe here the contents of the individual chapters, as this will be done at the beginning of the chapters themselves. Modular computations based on CRA are dealt with in Chapter 3, the Schönhage algorithm for multiplication being one of the topics. Chapter 4 develops CRA-based ideas in more extended contexts: polynomial interpolation, shift-register synthesis, cyclic convolution and fast Fourier transform, to name a few. Chapter 5 addresses the problem of carrying CRA-type arguments over to structures, where they would not be valid as such, that is, building a bridge from the "CRA-territory" to some parts outside of it.

Codes, the second of the C's in our landscape, form the subject matter of Chapter 6. The basic technique in coding theory is to add redundancy to data sent via a noisy channel or stored in a computer, this being done because of error detection and correction. In general, for modular algorithms based on CRT, the product of the moduli should be large enough with respect to the data. But it should not be too large, since this would be harmful to the efficiency of the modular algorithm. In applications to coding theory, however, matters look quite different, since making the product larger increases the possibilities of adding redundancy.

The third C, *cryptography*, is the topic of our final Chapter 7. The Chinese Remainder Theorem is in itself a *secret-sharing scheme*. Different parties are each in possession of a number. When they combine their knowledge, they find out a secret number, using CRT. This line of argument was clearly present already in the writing of Fibonacci quoted above, as well as in the actions of the ancient Chinese general who arranged his troops in rows of 9, 10, and 11 soldiers, by turns, and found out their exact number, without having to disclose it to anybody not knowing the CRT-technique.

The interdependence of the chapters can be roughly described by the following chart.

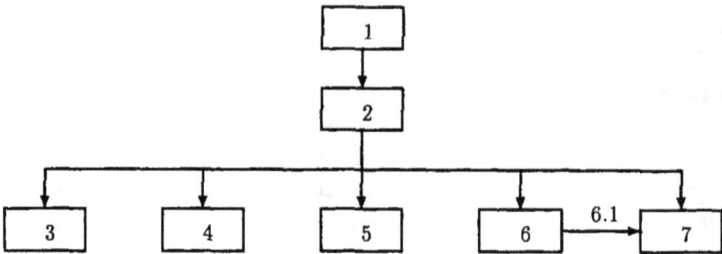

Figure 1.1: A flowchart of the interdependence of the chapters.

Chapter 2

Chinese Remainder Algorithm

In this chapter we shall first have a look at the motivations and historical development of the Chinese Remainder Theorem (CRT) and Algorithm (CRA). We then go into various kinds of generalizations of the Chinese Remainder Theorem and Algorithm, which are necessary for our applications in later chapters. It is interesting to note that modular computation was already known to ancient people, and that Chinese Remainder Algorithm was used in solving practical problems concerning the construction of walls and the base of buildings, the food trading, interests exchange, information delivery, and the computation of calendars in ancient times.

2.1 Historical Development

As early as the second century B.C. in ancient China congruences of first degree were applied for the calculation of calendars. This seems to be a main source of the remainder problem. In A.D. 237 the Chinese astronomers defined *shangyuan* as the starting point of the calendar in making *jingchu*. *Shangyuan* is a supposed moment that occurred simultaneously with the midnight of *Jiazi* (the first day of the 60 day cycle), the Winter Solstice and the new moon. If the Winter Solstice of a certain year occurred r_1 days after *shangyuan* and r_2 days after the new moon, then that year was N years after *shangyuan*. This results in the following system of congruences

$$aN = r_1 \bmod 60$$
$$aN = r_2 \bmod b,$$

where a is the number of days in a tropical year and b the number of days in a lunar month, and where the symbol "$x = y \bmod n$" means that x and y are congruent modulo n.

13

Sun Zi Suanjing is the mathematical classic of Sun Zi, a mathematician in ancient China. The exact date of this book is uncertain, however, according to Dickson's history book on number theory [29, p.57], it is about the first century A.D., see also the book *Niedere Zhalentheorie* (Leipzig, Teubner; Part 1, 1902, Part 2, 1910, i.83) by P. Bachman. The original Chinese Remainder Problem (CRP) proposed by Sun Zi in *Sun Zi Suanjing* (Problem 26, Volume 3), which consists of three volumes, is as follows:

> "We have a number of things, but do not know exactly how may. If we count them by threes we have two left over. If we count them by fives we have three left over. If we count them by sevens we have two left over. How many things are there?"

In modern terminology the problem is to find an x such that

$$x = 2 \bmod 3, \quad x = 3 \bmod 5, \quad x = 2 \bmod 7.$$

Sun Zi solved the problem as follows. The first step is to compute a value for the following s_0, s_1 and s_2:

$$s_0 = 0 \bmod 5 = 0 \bmod 7 = 1 \bmod 3,$$
$$s_1 = 0 \bmod 3 = 0 \bmod 7 = 1 \bmod 5,$$
$$s_2 = 0 \bmod 5 = 0 \bmod 3 = 1 \bmod 7.$$

He took $s_0 = 70, s_1 = 21$ and $s_2 = 15$. Since 5 and 7 divide s_0, s_0 must be of the form $7 \times 5 \times k = 35k$, where k is an integer. Hence $s_0 \bmod 3 = 2k \bmod 3$, and $k = 2$ gives $s_0 = 70$. s_1 and s_2 were similarly computed. The second step is to compute

$$s_0' = 2s_0 = 140, \quad s_1' = 3s_1 = 63, \quad s_2' = 2s_2 = 30.$$

The last step is to compute $x = s_0' + s_1' + s_2' \bmod (105) = 23$.

The solution was explained with a verse later in Cheng Dawei's book *Suanfa Tongzong* (A Collection of Algorithms, 1593). It reads as follows:

> "Three people walking together, it is rare that one be seventy. Five cherry blossom trees, twenty one branches bearing flowers, seven disciples reunite for the half-month. Taking away one hundred and five you shall know."

This verse means the following. Multiply by 70 the remainder of x when divided by 3, multiply by 21 the remainder of x when divided by 5, multiply by 15 the remainder of x when divided by 7. Add the three results together, and then subtract a suitable multiple of 105, and you shall have the required smallest solution.

A generalization of Sun's remainder problem is to find an integer x such that

$$x = r_i \bmod m_i, \quad i = 0, 1, ..., n - 1, \tag{2.1}$$

where m_i are pairwise relatively prime, and r_i are integers. Sun's algorithm for the general case is the following. First compute a value for $s_0, ..., s_{n-1}$ satisfying

$$s_i = 1 \bmod m_i, \quad s_i = 0 \bmod m_j \text{ for } i \neq j. \tag{2.2}$$

Then $x = \sum_{i=0}^{n-1} r_i s_i \bmod m$ is the smallest nonnegative solution of (2.1), where $m = m_0 m_1 \cdots m_{n-1}$ and the symbol "$a \bmod m$" denotes the smallest nonnegative integer congruent to a modulo m. It should be pointed out that Sun Zi did not mention a general method to solve (2.1). However, Sun Zi's method can be easily generalized, as done above. This algorithm is called the Chinese Remainder Algorithm (CRA).

Let $M_i = m/m_i$. To solve (2.2) is equivalent to solving the congruence

$$M_i x = 1 \bmod m_i, \tag{2.3}$$

i.e., computing the multiplicative inverse of M_i modulo m_i, where M_i and m_i are relatively prime. Furthermore, solving (2.3) is equivalent to finding two integers u_i and v_i such that

$$M_i u_i + m_i v_i = 1. \tag{2.4}$$

This diophantine equation can be solved by an algorithm, which is called Euclidean algorithm and was called *Gengxiang Jiansun* in ancient China in the book *Jiuzhang Suanshu* (Nine Chapters on the Mathematical Art). More generally, Euclidean algorithm can be used to find the greatest common divisor of two integers a and b, denoted by $\gcd(a, b)$. The algorithm goes as follows. Repeating the division, we obtain

$$
\begin{aligned}
a &= q_1 b + r_1, & 0 \leq r_1 < b, \\
b &= q_2 r_1 + r_2, & 0 \leq r_2 < r_1, \\
r_1 &= q_2 r_2 + r_3, & 0 \leq r_3 < r_2, \\
&\vdots & \vdots \\
r_{n-2} &= q_n r_{n-1} + r_n, & 0 \leq r_n < r_{n-1}, \\
r_{n-1} &= q_{n+1} r_n.
\end{aligned}
\tag{2.5}
$$

Since $b > r_1 > r_2 > \cdots \geq 0$, there must exist an integer n such that $r_{n+1} = 0$. Apparently,

$$\gcd(a, b) = \gcd(b, r_1) = \gcd(r_1, r_2) = \cdots = \gcd(r_{n-1}, r_n) = r_n.$$

Going backward in the above iterative procedure, we can express $\gcd(a, b) = r_n$ as a linear combination (with integral coefficients) of r_{n-1} and r_{n-2}, as a linear combination of r_{n-2} and r_{n-3}, ..., and finally a linear combination of a and b. This solves naturally (2.4) if we put $a = M_i$ and $b = m_i$. The Euclidean algorithm for polynomials of $F[x]$ works in the same way as that for integers.

As we observed, Sun Zi's problem and algorithm were given only by numerical examples without a general theory. Qin Jiushao (1202-1261) gave a general method to solve the Chinese remainder problems and systems of congruences of first degree in his book *Shushu Jiuzhang* (Nine Chapters of Mathematics, 1247). Gauss (1777-1855) completed the theory about congruences of first degree in his book *Disquisitiones Arithmeticae* (1801).

The moduli m_i were restricted to pairwise relatively prime numbers in the book *Sun Zi Suanjing*. Qin Jiushao treated moduli which are not necessarily relatively prime. He gave a method to convert the latter case into the former one in his book *Shushu Jiuzhang* [83]. Gauss studied the same conversion problem in his *Disquisitiones Arithmeticae*. By the fundamental theorem of arithmetic, Gauss factored each of the moduli into a product of prime powers, and then canceled all unnecessary congruences. For example, with Gauss' method the system of congruences

$$x = 17 \bmod 504,$$
$$x = -4 \bmod 35, \qquad\qquad (2.6)$$
$$x = 1 \bmod 16$$

is converted first into

$$x = 17 \bmod 9,$$
$$x = 17 \bmod 8,$$
$$x = 17 \bmod 7,$$
$$x = -4 \bmod 5,$$
$$x = -4 \bmod 7,$$
$$x = 1 \bmod 16.$$

Since $x = 1 \bmod 16$ implies that $x = 17 \bmod 8$, and $x = 17 \bmod 7$ and $x = -4 \bmod 7$ are the same, we should cancel the two congruences $x = 17 \bmod 8$ and $x = 17 \bmod 7$. This results in the following set of congruences

$$x = 17 \bmod 9,$$
$$x = -4 \bmod 7,$$
$$x = -4 \bmod 5,$$
$$x = 1 \bmod 16$$

with moduli pairwise relatively prime.

It is interesting that Qin solved the same problem without the concept of prime numbers 550 years before Gauss did. Qin's method is described as follows.

Let $m_1, m_2, ..., m_k$ be the moduli, and $l = \text{lcm}(m_1, ..., m_k)$, the least common multiple of $m_1, ..., m_k$. Qin's method is to find a set of integers $\mu_1, \mu_2, ..., \mu_k$ satisfying

1. μ_i divides m_i, $i = 1, 2, ..., k$;

2. $\gcd(\mu_i, \mu_j) = 1$ for all $i \neq j$;

3. $\mu_1 \mu_2 \cdots \mu_k = \text{lcm}(m_1, ..., m_k)$.

Then the system of congruences $x = a_i \bmod m_i$ for $i = 1, ..., k$, is converted into $x = a_i \bmod \mu_i$ for $i = 1, 2, ..., k$, where the moduli μ_i are pairwise relatively prime. Qin called the integers μ_i *dingshu*.

Consider first the case $k = 2$. The integers μ_1 and μ_2 can be computed with the following procedure:

St1: Let $\gcd(m_1, m_2) = d_1$. If $\gcd(m_1/d_1, m_2) = 1$, then take $\mu_1 = m_1/d_1$, $\mu_2 = m_2$.

St2: If $\gcd(m_1, m_2/d_1) = 1$, then take $\mu_1 = m_1$, $\mu_2 = m_2/d_1$.

St3: If $\gcd(m_1, m_2/d_1) = d_2 > 1$, then compute $d_3 = \gcd(m_1/d_2, m_2 d_2/d_1)$, where d_2 divides d_1 and d_3 divides d_1/d_2. If $d_3 = 1$, then take $\mu_1 = m_1/d_2, \mu_2 = m_2 d_2/d_1$, otherwise compute $d_4 = \gcd(m_1/d_2 d_3, m_2 d_2 d_3/d_1)$. Continue this process until there exists an integer s such that $d_{s+1} = 1$. Such an s must exist because $d_1 > d_2 > \cdots \geq 0$. Then take

$$\mu_1 = \frac{m_1}{d_2 d_3 \cdots d_s},$$
$$\mu_2 = \frac{m_2 d_2 d_3 \cdots d_s}{d_1}.$$

Thus, $\gcd(\mu_1, \mu_2) = 1$. We shall refer to this method as *Qin algorithm*. In fact, St1 is unnecessary.

For the case of k moduli, we apply the above Qin algorithm to m_k and m_{k-1} first, obtaining $\mu_k^{(1)}, \mu_{k-1}'$. Then applying the same algorithm to $\mu_k^{(1)}$ and m_{k-2} gives $\mu_k^{(2)}, \mu_{k-2}'$, continue this procedure and finally apply the algorithm to $\mu_k^{(k-2)}$ and m_1, obtaining μ_k, μ_1'. Then the integers

$$\mu_1', \mu_2', ..., \mu_{k-1}', \mu_k$$

satisfy $\gcd(\mu_k, \mu'_i) = 1$ for $i = 1, 2, ..., k - 1$ and

$$\text{lcm}(\mu_1, \mu_2, ..., \mu_k) = \text{lcm}(\mu'_1, \mu'_2, ..., \mu'_{k-1}, \mu_k) = \mu_k \text{lcm}(\mu'_1, \mu'_2, ..., \mu'_{k-1}).$$

So far we have reduced the case of k moduli into that of $k - 1$. Repeating this procedure gives the required *dingshu* $\mu_1, ..., \mu_k$.

To illustrate the above Qin algorithm for the general case, we take the system of congruences (2.6) as an example. With the above Qin algorithm the *dingshu* of 16, 35, and 504 are 16, 35, 9. The computation of the *dingshu* is depicted by Figure 2.1:

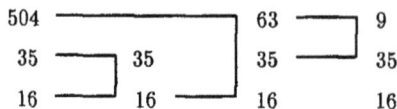

Figure 2.1: The procedure of Qin's algorithm.

Hence (2.6) is equivalent to the following system of congruences

$$\left\{ \begin{array}{rcl} x & = & 17 \bmod 9, \\ x & = & -4 \bmod 35, \\ x & = & 1 \bmod 16. \end{array} \right.$$

It is interesting to note that modular computation was employed in practical computations concerning the construction of the wall and base of a building as well as the counting of the number of soldiers in ancient China. The Great Wall is a magnificent work of ancient construction, built by the princes of the contending states in feudal times (475-221 B.C.). Behind the wall computation or cutting pipes is the following problem of computing the length L of the pipe, which was the representative of one of nine classes of practical remainder problems in the book *Shushu Jiuzhang* (1247) by Qin Jiushao.

The computation problem for building a wall or the base of a house proposed in the book *Shushu Jiuzhang* is described as follows. To construct a rectangular base for a building, there are four kinds of materials available: big cubic materials with each side 130 *Fen* long; small cubic materials with each side 110 *Fen*, city bricks with length 120 *Fen*, width 60 *Fen* and depth 25 *Fen*; and "six-door" bricks with length 100 *Fen*, width 50 *Fen* and depth 20 *Fen*. The craftsman is asked to use any kind of the four materials to finish the base without breaking the materials (it is possible to use each face of the bricks). After measuring, the craftsman found that with any kind of materials it is impossible to finish the base without breaking the materials. In detail, if big cubic materials are used, then 60-*Fen* base length is left, but 60-*Fen* more base width is needed. If small cubic materials are used, then 20-*Fen* base length is left, but 30-*Fen* more base width is needed. If the length of the city bricks is used, then 30-*Fen* base length is left, but 10-*Fen* more base width is needed. If the width of the city bricks is used, then 30-*Fen* base length is left, but 10-*Fen* more base width is needed. If the depth of the city bricks is used, then 25-*Fen* base length is left, and 10-*Fen* base width is left. If the length, width and depth of the six-door bricks are used, the base length has 30, 30 and 10 *Fens* left respectively, and the base width has 10, 10 and 10 Fens left respectively. The question is how large the base length N_1 and base width N_2 are.

This is a typical Chinese remainder problem which can be described in modern terminology as follows:

$$
\begin{aligned}
N_1 &= 60 \bmod 130 \\
    &= 30 \bmod 120 \\
    &= 20 \bmod 110 \\
    &= 30 \bmod 100 \\
    &= 30 \bmod 60 \\
    &= 30 \bmod 50 \\
    &= 5 \bmod 25 \\
    &= 10 \bmod 20 \\
N_2 &= 70 \bmod 130 \\
    &= 110 \bmod 120 \\
    &= 80 \bmod 110 \\
    &= 10 \bmod 100 \\
    &= 50 \bmod 60 \\
    &= 10 \bmod 50 \\
    &= 10 \bmod 25 \\
    &= 10 \bmod 20
\end{aligned}
$$

As we observed before, in ancient China Qin algorithm was employed to solve such a remainder problem.

It has been told that the CRA was employed to compute the number of solders in ancient times by some generals in China. A general asks his soldiers to stand in $m_1, m_2, ..., m_k$ rows in turn, and each time he counts the remainders. Finally, he computes the number of his solders with the CRA. In ancient China this was referred to as the *secret computation*, since at that time few people knew the technique. This was intended to prevent other people, especially the enemy, from knowing the number of his soldiers.

Apart from the calendar computation and the building problem as well as the counting of the number of soldiers, Qin Jioushao described a number of

other classes of remainder problems concerning food trading, interests exchange
and information delivering for military purposes.

As we saw, the solving of the congruence $ax = 1 \bmod b$ is essential to
the CRA. Qin also presented in his book *Shushu Jiuzhang* a method to solve
this congruence, which is called *Dayan Qiuyi Shu*. Suppose $0 < a < b$ and
$\gcd(a,b) = 1$. Applying the Euclidean algorithm to a and b as in (2.5), we
obtain $q_1 = 0, q_2, ..., q_n$ with $r_{n+1} = 0$. Define

$$j_0 = 0, \quad j_1 = 1, \quad j_s = q_s j_{s-1} + j_{s-2}, \quad s = 2, 3, ..., n.$$

Qin pointed out that $x = j_n = q_n j_{n-1} + j_{n-2}$ is the solution of $ax = 1 \bmod b$
if n is odd. When n is even, we need to do one more step of the algorithm
in (2.5), taking $r_{n-1} = q_{n+1} r_n + r'_{n+1}$, $r_n = q_{n+2} r'_{n+1}$ with $q_{n+1} = r_{n-1} - 1$,
$q_{n+2} = 1, r'_{n+1} = 1$. Hence $r_{n+2} = 0$ with $n + 1$ being odd. The *Dayan Qiuyi
Shu* can also be applied when n is even.

C. F. Gauss suggested also a method to solve the equation $ax + by = 1$,
where $\gcd(a,b) = 1$. If $A, B, C, D, ...$ and $\alpha, \beta, \gamma, \delta, ...$ have the relations

$$
\begin{aligned}
A &= \alpha, \\
B &= \beta A + 1, \\
C &= \gamma B + A, \\
D &= \delta C + B, ...
\end{aligned}
$$

then write

$$A = [\alpha], \quad B = [\alpha, \beta], \quad C = [\alpha, \beta, \gamma], \quad D = [\alpha, \beta, \gamma, \delta], ...,$$

where $[a_0, a_1, ..., a_n]$ denotes the finite continued fraction

$$a_0 + \cfrac{1}{a_1 + \cfrac{1}{a_2 + \cdots + \frac{1}{a_n}}}.$$

It then follows that

$$a = [q_{n+1}, q_n, ..., q_1], \quad b = [q_{n+1}, q_n, ..., q_2].$$

Putting

$$x = [q_n, q_{n-1}, ..., q_2], \quad y = [q_n, q_{n-1}, ..., q_1],$$

Gauss showed that $ax = by + 1$ if n is odd, and $ax = by - 1$ if n is even.

J. L. Lagrange (1736-1813) studied the same problem using continued fractions. He expressed the fraction a/b as

$$\frac{a}{b} = q_1 + \cfrac{1}{q_2 + \cfrac{1}{q_3 + \cdots + \frac{1}{q_{n+1}}}}.$$

Omitting the last term, and defining

$$\frac{y}{x} = q_1 + \cfrac{1}{q_2 + \cfrac{1}{q_3 + \cdots + \frac{1}{q_n}}},$$

Lagrange proved that $ax = by + 1$ if n is odd, and $ax = by - 1$ if n is even. It is not hard to see that the methods of Qin, Gauss and Lagrange are essentially the same.

In 1841 A. L. Crelle noted that $ax = 1 \bmod m$ has solution $a^{\Phi(m)-1}$, where $\Phi(m)$ is the Euler function [29, p.56].

The solution of the equation

$$ax + c = by$$

where a, b, c are given integers with $a > b$ and $\gcd(a, b) = 1$, is called *Kuttaka* by Indian mathematicians [4, 27, 89]. It was told that the Indian mathematician Aryabhata (c. A.D. 476) was able to solve the two-congruence Chinese remainder problem with the help of continued division (Euclidean algorithm) (see *Aryabhatiya* by Aryabhata, English translation by K. S. Shukla, 1979, Delhi).

According to [29, p.59] an Islamic scholar Ibn al-Haitam (about 1000) gave two methods to find a number, divisible by 7, which has the remainder 1 when divided by 2, 3, 4, 5, or 6. The first method gives the unique solution $1 + 2 \cdot 3 \cdot 4 \cdot 5 \cdot 6 = 721$. The second method gives a series of solutions 301, etc.

The Chinese remainder problem has been also considered by others throughout the world. According to Dickson [29, p.58], Nicomachus (about A.D. 100) gave the same problem and solution 23 as Sun Zi's. But Dickson's above remark was only based on a work published in 1866. Others who considered the Chinese remainder problems include Brahmegupta (A.D. 597), Bháscara (born, A.D. 1114), Gauss, Gegiomontanus (1436-1476), Hindenburg, Hutton, Kästner, Lagrange, Lüdicke, Misrachi (1455-1525), Pisano, Robinson, Thacker, Yih-hing (A.D. 717), etc.

According to [29, p.57], the Chinese remainder problem and algorithm became known in Europe through an article, "Jotting on the science of Chinese arithmetic," by Alexander Wylie (an English missionary), a part of which was translated into German by K. L. Biernatzki (see Jour. für Math., 52, 1856, 59-94), and into French by O. Terquem (see Nouv. Ann. Math., (2), 1, 1862, 35-44; 2, 1863, 529-540).

2.2   Chinese Remainder Algorithms

In the last section we have already described the Chinese Remainder Algorithm for integers. Another frequently employed version is the Chinese Remainder Algorithm for polynomials of $F[x]$, where F is a field. The Chinese Remainder Problem and Algorithm can also be generalized for Euclidean domains. In this section, we describe two versions of the Chinese Remainder Algorithm for any Euclidean domain.

There are two main algorithms which solve the n-congruence Chinese Remainder Problem (CRP) over an Euclidean domain D:

$$u = r_i \bmod m_i, \quad i = 0, 1, ..., n - 1, \tag{2.7}$$

where $r_i, m_i, u \in D$, and m_i are pairwise relatively prime. One is an iterative procedure based on an algorithm for the two-congruence CRP, and the other is a direct computation of the solution of the n-congruence CRP. We first describe the iterative version. To this end, we first treat the two-congruence CRP.

It is easily seen that the two-congruence CRP of (2.7) has a solution

$$u = r_0 + am_0,$$

where $a = (r_1 - r_0)b \bmod m_1$, and where $bm_0 = 1 \bmod m_1$, i.e., b is the inverse of m_0 modulo m_1. This gives naturally an algorithm for the two-congruence CRP as follows:

A two-congruence CRA

St0: Input m_0, m_1, r_0, r_1.

St1: Compute the inverse b of m_0 modulo m_1.

St2: $a := (r_1 - r_0) \times b$.

St3: Output $u := r_0 + am_0$.

To develop an algorithm for the n-congruence CRP, for $0 \le i \le n$ let

$$M_i = \begin{cases} 1, & \text{if } i = 0, \\ \prod_{k=0}^{i-1} m_i, & \text{otherwise.} \end{cases}$$

Since m_i are pairwise relatively prime, M_i and m_i are relatively prime. Suppose that u' is a solution of the k-congruence CRP, i.e.,

$$u' = r_i \bmod m_i, \quad i = 0, 1, ..., k - 1.$$

Let u be the solution of the following two-congruence CRP

$$u = u' \bmod M_k, \quad u = r_k \bmod m_k.$$

Then u is a solution of the k-congruence CRP. Thus, we can iterate the two-congruence CRA $n - 1$ times with respect to the pairs of moduli (M_i, m_i). This results in the following algorithm for the n-congruence CRP.

The n-congruence iterative CRA:

St0: Input m_i and r_i for $i = 0, 1, ..., n - 1$.

St1: Set $M := 1$, $U := r_0 \bmod m_0$.

St2: For $i := 1$ to $n - 1$ do the following.
$M := M \times m_{i-1}$;
$b :=$ the inverse of M modulo m_i;
$a := [(r_i - (U \bmod m_i)) \times b] \bmod m_i$;
$U := U + a \times M$.

St3: Output U.

The standard non-iterative CRA can be derived as follows. Let $M_k = \prod_{i=0, i\neq k}^{n-1} m_i$. Then M_k and m_k are relatively prime. It follows that there are two elements u_i and $v_i \in D$ such that $u_i M_i + v_i m_i = 1$. Then the solution to the n-congruence CRP of (2.7) is given by

$$u = \sum_{i=0}^{n-1} r_i u_i M_i \bmod M, \qquad (2.8)$$

where $M = m_0 m_1 \cdots m_{n-1}$. Thus, the non-iterative CRA can be described as follows.

The non-iterative CRA:

St0: Input n, m_i, r_i for $i = 0, 1, ..., n - 1$.

St1: Compute $M_k = \prod_{i=0, i\neq k}^{n-1} m_i$ for $k = 0, 1, ..., n - 1$.

St2: Compute the inverse u_i of M_i modulo m_i for $i = 0, 1, ..., n - 1$.

St3: Compute $u = \sum_{i=0}^{n-1} r_i u_i M_i \bmod M$.

St4: Output u.

It is not difficult to prove that the complexity of the above two CRAs is $O(n^2)$ with respect to the integer addition, subtraction, multiplication, and division. For most applications we need the CRAs for integers and for polynomials, i.e., in most applications the two CRAs for the Euclidean domain $\mathbf{Z}$ and $F[x]$ are of concern only when F is a field. There are also other variants of the above CRA. By CRA we refer to all algorithms solving the CRP.

2.3   Chinese Remainder Theorem

In Sections 2.1 and 2.2 we have seen that the Chinese Remainder Problem has at least one solution. In this section we prove that the solution is unique up to the product $m = m_0 m_1 \cdots m_{n-1}$, where m_i are the pairwise relatively prime moduli.

The Chinese Remainder Theorem (CRT) for integers can be stated as follows.

Theorem 2.3.1 *Let m_i be pairwise relatively prime. For any set of integers a_i the system of congruences*

$$x = a_i \bmod m_i, \quad i = 0, 1, ..., n-1$$

has exactly one solution modulo $m = m_0 m_1 \cdots m_{n-1}$.

Proof: The existence of solutions is proved by the Chinese Remainder Algorithms in Sections 2.1 and 2.2. If x and y are two solutions, then $x - y = 0 \bmod m_i$ for $i = 0, 1, ..., n-1$. It follows that $x = y \bmod m$. This proves the uniqueness. □

Similarly, one can prove the following Chinese Remainder Theorem for polynomials.

Theorem 2.3.2 *Let $m_i(x)$ be pairwise relatively prime polynomials of $F[x]$. Then for any set of polynomials $a_i(x)$ the following system of congruences*

$$u(x) = a_i(x) \bmod m_i(x), \quad i = 0, 1, ..., n-1$$

has exactly one solution modulo $m(x) = m_0(x) m_1(x) \cdots m_{n-1}(x)$.

The above two versions of the CRT are only of concern for our applications in later chapters. However, it could be helpful to interpret the CRT from a ring-theoretic point of view. If $R_1, R_2, ..., R_n$ are commutative rings with multiplicative identity, the set $R_1 \times R_2 \times \cdots \times R_n$ is a commutative ring with identity with respect to the following multiplication and addition componentwise

$$(a_1, a_2, ..., a_n)(b_1, ..., b_n) = (a_1 b_1, a_2 b_2, ..., a_n b_n),$$
$$(a_1, a_2, ..., a_n) + (b_1, ..., b_n) = (a_1 + b_1, a_2 + b_2, ..., a_n + b_n).$$

Let R be a commutative ring with identity, and I be an ideal of R. Then the mapping φ given by

$$\varphi_1 : r \in R \mapsto r + I$$

is called the natural homomorphism from R to the quotient ring R/I.

Theorem 2.3.3 *Let R be a commutative ring with identity. Suppose that $I_1, I_2, ..., I_n$ are ideals of R. Let $I = \cap_{i=1}^{n} I_i$. Then*

$$R/I \cong R/I_1 \times R/I_2 \times \cdots \times R/I_n,$$

where the isomorphism is given by

$$\varphi(r + I) = (r + I_1, r + I_2, ..., r + I_n).$$

Proof: Let $\varphi'(r) = (r + I_1, r + I_2, ..., r + I_n)$. It is easy to check that φ' is a homomorphism from R to $R/I_1 \times R/I_2 \times \cdots \times R/I_n$, and that the kernel $(\varphi')^{-1}(0, 0, ..., 0) = I$. Thus, the conclusions of the theorem follow.   □

This is the Chinese Remainder Theorem for commutative rings. The version of this theorem for integers is the following. Let m_i be positive integers, and $l = \text{lcm}(m_1, m_2, ..., m_n)$. Then

$$\mathbf{Z}/(l) \cong \mathbf{Z}/(m_1) \times \mathbf{Z}/(m_2) \times \cdots \times \mathbf{Z}/(m_n),$$

where the isomorphism is given by

$$\varphi(r + (l)) = (r + (m_1), r + (m_2), ..., r + (m_n)),$$

and where (a) denotes the ideal generated by a, and hereafter $\mathbf{Z}$ is the set of integers. One can also easily write down the version of this theorem for polynomials. These two versions of Theorem 2.3.3 will be useful for our applications in later chapters.

2.4   A Generalized CRA

In some applications we are interested in whether the system of congruences

$$x = a_i \bmod m_i, \quad i = 1, ..., k \tag{2.9}$$

has solutions, and how to compute them if there are some, where m_i are not pairwise relatively prime. We first consider the case $k = 2$ in which the following conclusion holds.

Theorem 2.4.1 *Let m be the least common multiple of two positive integers m_1 and m_2. Then the system of congruences*

$$x = a_1 \bmod m_1, \quad x = a_2 \bmod m_2 \tag{2.10}$$

has solutions if and only if

$$\gcd(m_1, m_2) | a_1 - a_2, \tag{2.11}$$

where $a|b$ means that a divides b. When the condition (2.11) holds, the system of (2.10) has only one solution modulo m.

Proof: Let $d = \gcd(m_1, m_2)$. If the system of (2.10) has a solution x_0, then

$$x_0 = a_1 \bmod d, \quad x_0 = a_2 \bmod d.$$

Hence $a_1 = a_2 \bmod d$, i.e., $d|a_1 - a_2$.

Now assume $d|a_1 - a_2$. Notice that any solution of the first congruence of (2.10) must be of the form

$$x = a_1 + m_1 y,$$

where y is some integer. Combining this with the second congruence of (2.10) gives

$$a_1 + m_1 y = a_2 \bmod m_2,$$

from which it follows that

$$\frac{m_1}{d} y = \frac{a_1 - a_2}{d} \bmod \frac{m_2}{d}. \tag{2.12}$$

Since m_1/d are m_2/d are relatively prime, using Euclidean algorithm we can find an integer y_0 such that $m_1 y_0/d = 1 \bmod (m_2/d)$. Thus, $y = y_0(a_2 - a_1)/d$ is a solution of (2.12). It is easy to see that the congruence (2.12) has only one solution modulo m_2/d. It then follows from $x = a_1 + m_1 y$ that the system (2.10) has only one solution modulo $m = m_1 m_2/d$. □

By induction on k Theorem 2.4.1 can be generalized into the following one.

Theorem 2.4.2 *Let $m_1, ..., m_k$ be positive integers. For a set of integers $a_1, ..., a_k$ the system of congruences (2.9) has solutions if and only if*

$$a_i = a_j \bmod \gcd(m_i, m_j), \quad i \neq j, \ 1 \leq i, j \leq k. \tag{2.13}$$

If conditions (2.13) hold, the solution is unique modulo $l = \mathrm{lcm}(m_1, ..., m_k)$.

In some applications we are also interested in the exact solutions of (2.9). A method for solving the system of congruences (2.9) was given by Ore [65], and can be described as follows. Let

$$m = \text{lcm}(m_1, m_2, ..., m_k), \quad B_i = m/m_i, \quad i = 1, ..., k.$$

Since $ab = \text{lcm}(a, b) \gcd(a, b)$, we obtain

$$
\begin{aligned}
B_i &= \text{lcm}\left(\frac{\text{lcm}(m_1, m_i)}{m_i}, \frac{\text{lcm}(m_2, m_i)}{m_i}, ..., \frac{\text{lcm}(m_k, m_i)}{m_i}\right) \\
&= \text{lcm}\left(\frac{m_1}{\gcd(m_1, m_i)}, \frac{m_2}{\gcd(m_1, m_i)}, ..., \frac{m_k}{\gcd(m_k, m_i)}\right) \\
&= \text{lcm}\left(\frac{m_1}{d_{1i}}, \frac{m_1}{d_{2i}}, ..., \frac{m_1}{d_{ki}}\right),
\end{aligned}
\tag{2.14}
$$

where $d_{ij} = \gcd(m_i, m_j)$.

Theorem 2.4.3 *The solution of the simultaneous congruences (2.9) is given by*

$$x = a_1 c_1 \frac{m}{m_1} + a_2 c_2 \frac{m}{m_2} + \cdots + a_k c_k \frac{m}{m_k} \bmod m, \tag{2.15}$$

where the c_i form a set of integers satisfying the condition

$$c_1 \frac{m}{m_1} + c_2 \frac{m}{m_2} + \cdots + c_k \frac{m}{m_k} = 1. \tag{2.16}$$

Proof: We first show that $\gcd(B_1, B_2, ..., B_k) = 1$. It suffices to show that for each prime p at least one of the integers B_i is not divisible by p. Let p^{α_i} be the highest power of p dividing m_i, and suppose without loss of generality that α_1 is the greatest of these exponents, then m is divisible by p^{α_1}, but B_1 is not divisible by p.

Applying Euclidean algorithm to B_1 and B_2, we can find integers α_1 and α_2 such that $\alpha_1 B_1 + \alpha_2 B_2 = \gcd(B_1, B_2)$. Then an application of Euclidean algorithm to $\gcd(B_1, B_2)$ and B_3 yields integers $\beta_1, \beta_2, \beta_3$ such that $\beta_1 B_1 + \beta_2 B_2 + \beta_3 B_3 = \gcd(B_1, B_2, B_3)$. Repeating this process finally yields integers c_i such that (2.16) holds.

We now prove that the x given by (2.15) is a solution to (2.9). Multiplying congruence (2.13) by B_i gives

$$B_i a_i \equiv B_i a_j \bmod d_{ij} B_i.$$

But according to (2.14) the integer B_i is divisible by m_i/d_{ij}. Hence

$$B_i a_i \equiv B_i a_j \bmod m_j.$$

For each i we multiply this congruence by c_i. The summation of these congruences yields

$$x = a_j(c_1 B_1 + c_2 B_2 + \cdots c_k B_k) \bmod m_j.$$

By (2.16) we have $x = a_j \bmod m_j$. This completes the proof.   □

Theorem 2.4.3 gives the following algorithm, which is a generalization of the non-iterative Chinese Remainder Algorithm. The generalization is denoted by GCRA.

Generalized CRA:

St0: Input $m_1, ..., m_k, a_1, ..., a_k$.

St1: Set $m =: m_1$.

St2: For $i = 1$ to $k - 1$ do the following

$$a =: \gcd(m, m_{i+1}), \quad m =: mm_{i+1}/a.$$

St3: Output m.

St4: Compute $B_i = m/m_i$ for $i = 1, ..., k$.

St5: Apply extended Euclidean algorithm to B_i to find integers c_i such that $c_1 B_1 + c_2 B_2 + \cdots + c_k B_k = 1$.

St6: Compute $x = \sum_{i=1}^{k} a_i c_i B_i \bmod m$.

St7: Output x.

We consider now an example to illustrate Theorem 2.4.3 and the corresponding algorithm above. Consider the three moduli $m_1 = 4, m_2 = 6, m_3 = 22$. We have

$$m = \operatorname{lcm}\{4, 6, 22\} = \operatorname{lcm}\{\operatorname{lcm}\{4, 6\}, 22\} = \operatorname{lcm}\{12, 22\} = 132.$$

First we apply Euclidean algorithm to $m/m_1 = 33$ and $m/m_2 = 22$, obtaining

$$11 = \gcd(33, 22) = 33 - 22.$$

Then apply it to 11 and $m/m_3 = 6$, obtaining

$$1 = 2 \times 6 - 11 = (-1) \times 33 + 1 \times 22 + 2 \times 6.$$

Hence if the system of equations

$$x = a_1 \bmod 4, \quad x = a_2 \bmod 6, \quad x = a_3 \bmod 22$$

has a solution modulo 132, it must be given by

$$x = -33a_1 + 22a_2 + 12a_3 \bmod 132.$$

Another approach to solving the system of congruences (2.9) is to convert it into one whose moduli are pairwise relatively prime, as described in Section 2.1.

2.5   Another Generalized CRT

In this section we introduce another generalization of the Chinese Remainder Theorem due to Fiol [33].

Let $\mathbf{Z}^n$ be the additive group consisting of vectors $\mathbf{a} = (a_1, ..., a_n)^T$, where $a_i \in \mathbf{Z}$. Let M be an $n \times n$ integral matrix with linearly independent columns, that is, $\det M \neq 0$. The set $M\mathbf{Z}^n$ whose elements are a linear combination (with integral coefficients) of the columns of M is said to be the lattice generated by M. Clearly, $M\mathbf{Z}^n$ is a normal subgroup of $\mathbf{Z}^n$. The quotient group $\mathbf{Z}^n/M\mathbf{Z}^n$ has order $m = \det M$.

Let $\mathbf{a} = (a_1, a_2, ..., a_n)^T, \mathbf{b} = (b_1, b_2, ..., b_n)^T \in \mathbf{Z}^n$. We say that $\mathbf{a}$ and $\mathbf{b}$ are congruent modulo M and write $\mathbf{a} = \mathbf{b} \bmod M$ if $\mathbf{a} - \mathbf{b} \in M\mathbf{Z}^n$. In particular, when $M = \mathrm{diag}(m_1, ..., m_m)$, a and b are congruent modulo M if and only if the system of congruences

$$a_i = b_i \bmod m_i, \quad i = 1, 2, \cdots, n$$

holds. For an integral matrix M, it is well-known that there exist two unimodular (with determinant ± 1) integral matrices U and V such that

$$S = UMV = \mathrm{diag}(s_1, s_2, ..., s_n)$$

with $s_i | s_{i+1}$ for $i = 1, 2, ..., n - 1$. (For a proof, see [42]). Hence $\mathbf{a} = \mathbf{b} \bmod M$ if and only if

$$\mathbf{u}_i \mathbf{a}^T = \mathbf{u}_i \mathbf{b}^T \bmod s_i, \quad i = 1, 2, \cdots, n,$$

where $\mathbf{u}_i$ stands for the ith row of U.

For an element $\mathbf{a}$ of $\mathbf{Z}^n/M\mathbf{Z}^n$, let $o(\mathbf{a})$ denote its order which is defined to be the smallest positive integer r such that $r\mathbf{a} = 0 \bmod M$, or equivalently

$$rM^{-1}\mathbf{a} = \frac{|m|M^{-1}\mathbf{a}}{|m|/r} \in \mathbf{Z}^n.$$

Since r is minimal, $|m|/r$ is the greatest positive integer that divides m and all the numbers $\mathbf{m}_i \mathbf{a}^T$, where $\mathbf{m}_i$ denotes the ith row of mM^{-1}. Thus,

$$o(\mathbf{a}) = \frac{|m|}{g}, \tag{2.17}$$

where $g = \gcd(m, \mathbf{m}_1 \mathbf{a}^T, \mathbf{m}_2 \mathbf{a}^T, ..., \mathbf{m}_n \mathbf{a}^T)$.

The Chinese Remainder Theorem described by Theorem 2.3.1 has a slightly generalized version as follows.

Theorem 2.5.1 *Let $b_1, ..., b_n$ be arbitrary integers. Let $m_1, m_2, ..., m_n$ and a_1, $a_2, ..., a_n$ be integers such that*

$$\begin{aligned} \gcd(m_i, m_j) &= 1, \ i \neq j; \\ \gcd(a_i, m_i) &= 1, \ i = 1, 2, ..., n. \end{aligned} \tag{2.18}$$

Then the system of congruences

$$a_i x = b_i \bmod m_i, \quad i = 1, 2, \cdots, n \tag{2.19}$$

has exactly one solution modulo $m = m_1 m_2 \cdots m_n$.

Since $\gcd(a_i, m_i) = 1$, by Euclidean algorithm we can find integers c_i such that $c_i a_i = 1 \bmod m_i$ for each i. Therefore the system of congruences (2.19) is equivalent to the system of congruences

$$x = c_i a_i \bmod m_i, \quad i = 1, 2, \cdots, n,$$

which is the case described in Theorem 2.3.1.

The system of congruences (2.19) is equivalent to the congruence

$$x\mathbf{a} = \mathbf{b} \bmod M, \tag{2.20}$$

where $\mathbf{a} = (a_1, a_2, ..., a_n)^T$, $\mathbf{b} = (b_1, b_2, ..., b_n)^T$, and $M = \operatorname{diag}(m_1, m_2, ..., m_n)$. We are interested in the solution of the congruence (2.20) for ordinary matrices M.

Theorem 2.5.2 *Given an $n \times n$ integral matrix M with $m = \det M \neq 0$, if $g = 1$, then the congruence $x\mathbf{a} = b \bmod M$ has exactly one solution modulo m for any $\mathbf{b} \in \mathbf{Z}^n$, where g is defined as before.*

Proof: Suppose $g = 1$, we have $o(\mathbf{a}) = |m|$ by (2.17). The group $\mathbf{Z}^n/M\mathbf{Z}^n$ is cyclically generated by $\mathbf{a}$. Hence the congruence $x\mathbf{a} = \mathbf{b} \bmod M$ has solutions for any $\mathbf{b}$. Suppose that x_1 and x_2 are two solutions, then $(x_1 - x_2)\mathbf{a} = \mathbf{0} \bmod M$, from which it follows that $x_1 = x_2 \bmod m$, since $o(\mathbf{a}) = |m|$.   □

In particular, if $M = \operatorname{diag}(m_1, m_2, ..., m_n)$ then

$$g = \gcd\left(m, \frac{a_1 m}{m_1}, ..., \frac{a_n m}{m_n}\right) = 1,$$

which is equivalent to (2.18).

Similarly, we have the following generalization of the CRT for polynomials.

Theorem 2.5.3 *Let $b_1(x), ..., b_n(x)$ be arbitrary polynomials of $F[x]$, $m_1(x), ..., m_n(x)$ and $a_1(x), ..., a_n(x)$ be polynomials of $F[x]$ such that*

$$\gcd(m_i(x), m_j(x)) = 1, \quad i \neq j;$$
$$\gcd(a_i(x), m_i(x)) = 1, \quad i = 1, 2, ..., n.$$

Then the system of congruences

$$a_i(x)u(x) = b_i(x) \bmod m_i(x), \quad i = 1, 2, ..., n \tag{2.21}$$

has exactly one solution modulo $m(x) = m_1(x)m_2(x) \cdots m_n(x)$, where F is a field.

Let symbols be the same as in Theorem 2.5.3. The system of congruences of (2.21) is solved by the usual CRA for polynomials as follows. Since $\gcd(a_i(x), m_i(x)) = 1$, with Euclidean algorithm we compute a polynomial $c_i(x) \in F[x]$ such that

$$c_i(x)a_i(x) = 1 \bmod m_i(x)$$

for all i. Then (2.21) is converted into

$$u(x) = c_i(x)b_i(x) \bmod m_i(x), \quad i = 1, 2, \cdots, n,$$

which is solved by the usual CRA for polynomials.

We now take one example to illustrate Theorem 2.5.3. Consider the two moduli $m_1(x) = x^3 + x + 1$ and $m_2(x) = x^3 + x^2 + 1$ over $GF(2)$. Let

$$a_1(x) = x^2 + x + 1, \quad a_2(x) = x + 1.$$

It is easily seen that a_i and m_i are relatively prime for each i. With Euclidean algorithm we have

$$1 = (x + 1)m_1(x) + x^2 a_1(x)$$

and

$$1 = m_2(x) + x^2 a_2(x).$$

Thus, the inverses $c_i(x)$ of $a_i(x)$ are

$$c_1(x) = x^2, \quad c_2(x) = x^2.$$

Note that

$$x m_1(x) + (x + 1) m_2(x) = 1.$$

It follows that the solution of the system of congruences

$$\begin{aligned} a_1(x)u(x) &= r_1(x) \bmod m_1(x), \\ a_2(x)u(x) &= r_2(x) \bmod m_2(x) \end{aligned}$$

is given by

$$u(x) = (x + 1)x^2 m_2(x) r_1(x) + x^3 m_1(x) r_2(x) \bmod m(x).$$

where $m(x) = m_1(x) m_2(x)$.

We mention, finally, that these generalized CRTs and CRAs are necessary for our applications in constructing redundant residue codes and secret-sharing in later chapters. In addition, Theorem 2.5.2 may be useful in some applications when the matrix M is chosen of other special forms. There are also analogues of Theorem 2.5.2 for polynomials over arbitrary fields.

Chapter 3

In Modular Computations

The Chinese Remainder Problem arose originally from the computation of calendars. As we saw in Chapter 2, modular computation was employed in solving many other practical problems in ancient times. Gradually, this method has been generalized into the so-called modular technique and further extended into a more general kind of computing technique—the homomorphic image computing. This chapter is intended to summarize the essentials of modular and homomorphic image computing, to show the procedure of developing a modular algorithm, and to introduce some of the modular algorithms based on the CRA.

3.1 Modular Computation Based on CRA

Let $m_1, ..., m_k$ be k pairwise relatively prime integers with $m_i \geq 2$, and $m = \prod_{i=1}^{k} m_i$. Recall that the CRT shows that the mapping

$$\varphi(x \bmod m) = (x \bmod m_1, ..., x \bmod m_k)$$

is an isomorphism between the two rings $\mathbf{Z}/(m)$ and $\mathbf{Z}/(m_1) \times \cdots \times \mathbf{Z}/(m_k)$. For each integer x between 0 and $m - 1$ the representation

$$x = (x \bmod m_1, ..., x \bmod m_k)$$

is called the *Sino correspondence* or *Sino representation* in digital signal processing [39], and *modular representation* in computer arithmetic [46].

The basic idea of modular techniques based on CRA can be described as follows. Let $x^{(1)}, ..., x^{(t)}$ be elements of $\mathbf{Z}$ (resp. $\mathbf{Z}[x]$), and $F(x^{(1)}, ..., x^{(t)})$ be an arithmetic expression over $\mathbf{Z}$ (resp. $\mathbf{Z}[x]$) with operations $\{+, -, \times\}$ of $\mathbf{Z}$ (resp. $\mathbf{Z}[x]$). Let m be a positive integer and ϕ_m denote the ring homomorphism

$$\phi_m(y) = y \bmod m, \quad y \in \mathbf{Z} \text{ (resp. } \mathbf{Z}[x]).$$

33

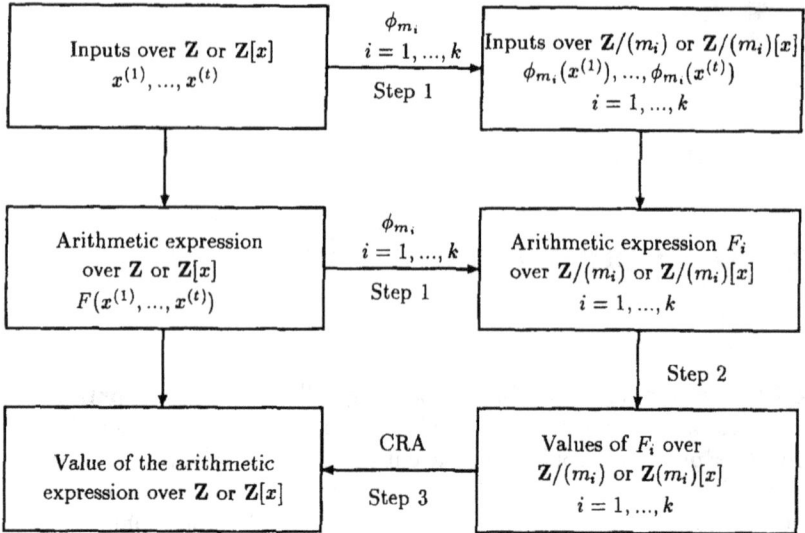

Figure 3.1: The idea of modular techniques based on the CRA

Instead of computing the $F(x^{(1)}, ..., x^{(t)})$ directly, the modular approach is first to choose a number of pairwise relatively prime moduli $m_1...., m_k$. Then for each m_i compute $\phi_{m_i}(x^j)$ for $j = 1, ..., t$ and the new arithmetic expression F_i over $\mathbf{Z}/(m_i)$ whose coefficients are the corresponding ones of F modulo m_i.

The second step of the modular approach is to compute the new arithmetic expressions $F_i(\phi_{m_i}(x^{(1)}), ..., \phi_{m_i}(x^{(t)}))$ for $i = 1, ..., k$. The third step is to apply the CRA or one of its variants to the images F_i to get the value F of the arithmetic expression $F(x^{(1)}, ..., x^{(t)})$. The idea of modular techniques based on CRA may be depicted by Figure 3.1.

One problem of modular techniques is the choice of the moduli, which depend on the arithmetic expression. From the viewpoint of computational complexity the moduli and the number of moduli should be as small as possible. But in order to get enough information about the value of the arithmetic expression $F(x^{(1)}, ..., x^{(t)})$ from the values of the new arithmetic expressions, the product of the moduli should be large enough. To illustrate the idea of modular techniques and the choices of the moduli, we take the arithmetic expression

$$F(x, y) = 100xy + x^2 + y^2$$

as an example. Suppose we want to calculate $F(x, y)$ for the inputs x and y in the ranges $0 \leq x \leq 10$ and $0 \leq y \leq 10$. Then for the inputs in the ranges we have

$$0 \leq F(x,y) \leq 10200.$$

It follows from the CRT that to compute the value of $F(x,y)$ for each pair of inputs (x,y) in the above ranges correctly it is necessary and sufficient to choose the pairwise relatively prime moduli $m_1, ..., m_k$ such that

$$m := \prod_{i=1}^{k} m_i \geq 10201.$$

Under this condition we choose $m_1 = 3, m_2 = 4, m_3 = 7, m_4 = 11$, and $m_5 = 13$. Then $m = 12012 > 10201$. To compute $F(10,9)$, we first perform Step 1 described in Figure 3.1. It is easily calculated

$$
\begin{aligned}
F_1(x,y) &= (x+y)^2 - xy, & (x,y) &\in \mathbf{Z}/(3) \times \mathbf{Z}/(3), \\
F_2(x,y) &= x^2 + y^2, & (x,y) &\in \mathbf{Z}/(4) \times \mathbf{Z}/(4), \\
F_3(x,y) &= (x+y)^2, & (x,y) &\in \mathbf{Z}/(7) \times \mathbf{Z}/(7), \\
F_4(x,y) &= (x+y)^2 - xy, & (x,y) &\in \mathbf{Z}/(11) \times \mathbf{Z}/(11), \\
F_5(x,y) &= (x-y)^2 - 2xy, & (x,y) &\in \mathbf{Z}/(13) \times \mathbf{Z}/(13)
\end{aligned}
$$

and

$$
\begin{aligned}
\phi_1(10,9) &= (1,0), & \phi_2(10,9) &= (2,1), \\
\phi_3(10,9) &= (3,2), & \phi_4(10,9) &= (-1,-2), & \phi_5(10,9) &= (10,9).
\end{aligned}
$$

Then the calculations of Step 2 show

$$F_1(1,0) = 1, \ F_2(2,1) = 1, \ F_3(3,2) = 4, \ F_4(-1,-2) = 7, \ F_5(10,9) = -1.$$

Finally, Step 3 is the application of the CRA or one of its variants by which we get $F(10,9) = 9181$.

Let the inputs $x^{(1)}, ..., x^{(t)}$ be confined to $a_i \leq x^{(i)} \leq b_i$ for each $1 \leq i \leq t$ and let

$$
\begin{aligned}
M_1 &= \max_{a_i \leq x^{(i)} \leq b_i} F(x^{(1)}, ..., x^{(t)}) - \min_{a_i \leq x^{(i)} \leq b_i} F(x^{(1)}, ..., x^{(t)}), \\
M_2 &= \max_{a_i \leq x^{(i)} \leq b_i} |F(x^{(1)}, ..., x^{(t)})|.
\end{aligned}
\qquad (3.1)
$$

To ensure $[F(x^{(1)}, ..., x^{(t)}) \bmod m] = F(x^{(1)}, ..., x^{(t)})$ for each input $(x^{(1)}, ..., x^{(t)})$ with $a_i \leq x^{(i)} \leq b_i$, it is necessary that $m > M_2 + 1$. For example, let $F(x) = x$, where $0 \leq x \leq b$, then $M_2 = b$ with respect to this arithmetic expression. If we choose $m = M_2$, then $F(b) \bmod m = F(0) \bmod m = 0$, but $F(b) = b \neq 0$. However, the condition $m > M_2 + 1$ is not enough to ensure that different

$F(x^{(1)}, ..., x^{(t)})$ have different images. For example, let $F(x) = x$, where $-b \leq x \leq b$, then $M_2 = b$. Choose $m = b+1$, then $F(b) \bmod m = F(-1) \bmod m = b$, but $b \neq -1$. Thus, it is necessary to require that $m > M_1 + 1$. It is easy to see that, to compute the value of the arithmetic expression $F(x^{(1)}, ..., x^{(t)})$ over $\mathbf{Z}$ correctly for each input $(x^{(1)}, ..., x^{(t)})$ with $a_i \leq x^{(i)} \leq b_i$ for $1 \leq i \leq t$, it is necessary and sufficient to choose the pairwise relatively prime moduli $m_1, ..., m_k$ such that

$$m := \prod_{i=1}^{k} m_i \geq \max\{M_1 + 1, M_2 + 1\}. \tag{3.2}$$

To illustrate the choice of the moduli, we take another example. Let

$$F(x, y) = 3x^3 + 4y^3, \quad |x| \leq 4, \quad |y| \leq 4,$$

be the arithmetic expression. It is easily seen that

$$M_1 = \max F(x, y) - \min F(x, y) = 448 + 448 = 896,$$
$$M_2 = 448.$$

Thus, the moduli should be chosen such that $m = \sum_{i=1}^{k} m_i > 896 + 1 = 897$. Since $897 = 3 \times 13 \times 23$, the set of moduli can be chosen as $m_1 = 3, m_2 = 13, m_3 = 23$. A better set of moduli may be $m_1 = 3, m_2 = 4, m_3 = 25$, since with this set of moduli we have

$$\begin{cases} F_1(x, y) = F(x, y) \bmod 3 = y^3, \\ F_2(x, y) = F(x, y) \bmod 4 = 3x^3, \\ F_3(x, y) = F(x, y) \bmod 25 = 3x^3 + 4y^3. \end{cases} \tag{3.3}$$

But if we choose the former set of moduli, we have

$$\begin{cases} F_1(x, y) = F(x, y) \bmod 3 = y^3, \\ F_2(x, y) = F(x, y) \bmod 13 = 3x^3 + 4y^3, \\ F_3(x, y) = F(x, y) \bmod 23 = 3x^3 + 4y^3, \end{cases}$$

which contain more multiplications and additions than those of (3.3).

Generally, the moduli must be chosen such that (3.2) holds. Under this condition there are many kinds of choices. In principle, m should be chosen as small as possible, but in some cases choosing a larger m could be beneficial, as shown by the choices of the moduli for the Schönhage algorithm in Section 3.2. Also one has to make a compromise between the sizes of m_i and the number k of moduli.

An important problem of modular algorithms based on CRA is the estimation of the M_1 and M_2 of (3.1). Using any two numbers larger than M_1 and

M_2 respectively in place of M_1 and M_2 works for a modular algorithm, but the tightness of lower bounds on M_1 and M_2 contributes to the performance of the algorithm.

For an arithmetic expression $F(y^{(1)}, ..., y^{(t)})$ over $\mathbf{Z}[x]$, where $y^{(i)} \in \mathbf{Z}[x]$, suppose that we want to compute the result of $F(y^{(1)}, ..., y^{(t)})$, where $y^{(i)} \in S_i \subseteq \mathbf{Z}[x]$. Let

$$D = \max_{y^{(i)} \in S_i} \deg F(y^{(1)}, ..., y^{(t)}).$$

Then the moduli $m_i(x) \in \mathbf{Z}[x]$ must be chosen such that

$$d := \deg \prod_{i=1}^{k} m_i(x) > D,$$

where m_i are pairwise relatively prime.

Consider now the following example. Let $F(y^{(1)}, y^{(2)}) = y^{(1)} y^{(2)}$, where $0 \leq \deg(y^{(1)}) \leq m, 0 \leq \deg(y^{(2)}) \leq n$. Then $D = m + n$. Thus, the moduli must be chosen such that

$$d := \deg \prod_{i=1}^{k} m_i(x) > m + n.$$

The most important question concerning modular techniques may be whether they are computationally worthwhile, since conversions into and out of modular representations take time. The first step within modular techniques based on the CRA includes the modularization of inputs, i.e., the calculation of $x \bmod m_i$ and the computation of the arithmetic expressions F_i for $i = 1, ..., k$. These are easily done by means of division, since the moduli are usually chosen small enough. The second step is the calculation of F_i for $i = 1, ..., k$. The primitive computations within this step are the addition, subtraction, and multiplication of $\mathbf{Z}/(m_1) \times \cdots \times \mathbf{Z}/(m_k)$, which can be carried out componentwise as follows:

$$(x_1, ..., x_k) \pm (y_1, ..., y_k) = ((x_1 \pm y_1) \bmod m_1, ..., (x_k \pm y_k) \bmod m_k),$$
$$(x_1, ..., x_k) \times (y_1, ..., y_k) = ((x_1 \times y_1) \bmod m_1, ..., (x_k \times y_k) \bmod m_k).$$

Thus, the modular approach reduces the calculation of an arithmetic expression to the addition, subtraction, and multiplication of the residue rings $\mathbf{Z}/(m_i)$. If the moduli are not very large, traditional algorithms are expected to execute these operations with a reasonable speed. If the moduli are of special forms, these operations can be carried out with proper expressions (see Section 3.2). As far as the order of complexity is concerned, modular approaches to addition and

subtraction have no advantage, since the order of the computational complexity
of adding or subtracting two n-digit numbers is essentially $O(n)$. However,
modular techniques based on the CRA make parallel computations possible.
Thus, these approaches to addition and subtraction do have a substantial gain
in speed if parallel mechanisms are installed.

One obvious advantage of modular approaches is the gain of speed in multiplication. It is well-known that the complexity of multiplying two n-digit
numbers is essentially of order n^2, though there are some algorithms which can
do a little better (see Section 3.2). Thus, the multiplication of $\mathbf{Z}/(m)$ has a
complexity of order

$$(\log m)^2 = \sum_{i=1}^{k}(\log m_i)^2 + \sum_{i \neq j}(\log m_i)(\log m_j),$$

whereas the multiplication of $\mathbf{Z}/(m_1) \times \cdots \times \mathbf{Z}/(m_k)$ has a complexity of order

$$\sum_{i=1}^{k}(\log m_i)^2.$$

This clearly shows one advantage of modular approaches. Thus, if an arithmetic
expression involves many multiplications, the gain within Step 2 should be much
greater than resources needed to compensate the computations in Steps 1 and
3. The third step is an application of the CRA whose complexity is of order
$O(k^2)$.

One disadvantage of a modular representation is that it is comparatively
difficult to test whether a number is positive or negative. In addition, it is comparatively difficult to test whether or not $(x_1, ..., x_k)$ is greater than $(y_1, ..., y_k)$.
It is also difficult to test whether or not overflow has occurred as the result of
addition, subtraction, or multiplication. However, these disadvantages are only
relevant to some applications.

It is clear that modular techniques based on CRA are divide-and-conquer
ones. Step 1 is the divide procedure, while Steps 2 and 3 constitute the conquer
procedure.

A more general extension of this technique is the so-called *homomorphic
image computing*, which will be outlined in Section 3.5.

3.2   A Modular Approach to Multiplication

The multiplication of integers is a primitive operation for many applications.
The first method we learned may be the *school algorithm*, which can be illustrated by the example 1234×321 in Figure 3.2. It is obvious that the school

```
        1  2  3  4
  ×           3  2  1
  ─────────────────────
        1  2  3  4
     2  4  6  8
  +  3  7  0  2
  ─────────────────────
     3  9  6  1  1  4
```

Figure 3.2: The school algorithm

algorithm has the complexity $O(mn)$ for the multiplication of an m-digit number and an n-digit number.

There is another method which has about the same complexity. The theoretical basis of the algorithm is quite simple. For any two positive integers a and b, it is easy to see that

$$a \times b = \left\lfloor \frac{a}{2} \right\rfloor \times 2b + \frac{1 - (-1)^a}{2} b.$$

Setting $a_0 = a$, $b_0 = b$, $a_i = \lfloor a_{i-1}/2 \rfloor$ and $b_i = 2b_{i-1}$, we have

$$
\begin{aligned}
a_i \times b_i &= a_{i+1} \times b_{i+1} + \frac{1 - (-1)^{a_i}}{2} b_i \\
&= a_{i+1} \times b_{i+1} + \begin{cases} 0, & \text{if } a_i \text{ even} \\ b_i, & \text{otherwise.} \end{cases}
\end{aligned}
$$

Thus, the algorithm can be described as follows. Write the multiplier and the multiplicand side by side. Then make two columns, one under each operand, by repeating the following rules until the number under the multiplier is 1:

- Divide the number under the multiplier by 2, ignoring any fractions;

- then double the number under the multiplicand by adding it to itself;

- finally, cross out each row in which the number under the multiplier is even and then add up the numbers that remain in the column under the multiplicand.

It is not difficult to prove that the complexity of this algorithm is $O(mn)$ for the multiplication of an m-bit number and an n-bit number. However, the algorithm may be a little better than the school algorithm, since the hidden constant could be smaller than that in the school algorithm.

For the multiplication of two n-digit numbers the two algorithms just described have the complexity $O(n^2)$. There is a divide-and-conquer algorithm which is faster than the above two algorithms. It has the complexity $n^{\log_2 3}$, and can be outlined as follows.

Let u and v be two integers of n decimal digits. Separate each of these operands into two parts as nearly the same size as possible:

$$u = 10^s w + x, \quad v = 10^s y + z,$$

where $0 \leq x < 10^s$, $0 \leq z < 10^s$, and $s = \lfloor n/2 \rfloor$. The integers w and y therefore both have $\lceil n/2 \rceil$ digits. Then the product becomes

$$uv = 10^{2s} wy + 10^s (wz + xy) + xz.$$

The trick of the divide-and-conquer algorithm is the introduction of the auxiliary variable

$$r = (w + x)(y + z) = wy + (wz + xy) + xz,$$

With the help of this variable the product is converted into

$$uv = 10^{2s} wy + 10^s (r - wy - xz) + xz.$$

Thus, to compute uv, only three multiplications of two integers having $\lceil n/2 \rceil$ digits are needed. Thus, a recursive algorithm for the multiplication of two large integers u and v, denoted by $M(u,v)$, can be described as follows:

Step 1: If n is small, then multiply u by v with the school algorithm; otherwise, go to Step 2.

Step 2: $s := n$ div 2
$w := u$ div 10^s, $x := u$ mod 10^s,
$y := v$ div 10^s, $z := v$ mod 10^s.

Step 3: Return

$$M(w,y) \times 10^{2s} + (M(w+x,y+z) - M(w,y) - M(x,z)) \times 10^s \atop + M(x,z). \quad (3.4)$$

Let $t(n)$ be the time required by the recursive algorithm to multiply two integers of size at most n. Taking account of the fact that $w + x$ and $y + z$ may be up to $1 + \lceil n/2 \rceil$ digits, we have

$$t(n) \leq t(\lfloor n/2 \rfloor) + t(\lceil n/2 \rceil) + t(1 + \lceil n/2 \rceil) + cn$$

for every $n \geq n_0$. This gives $t(n) \in O(n^{\log_2 3})$.

It should be noted that the hidden constant may be quite large, so the algorithm becomes interesting in practice only when n is quite large. It is also worth mentioning that divide-and-conquer approaches do not necessarily result in a better algorithm. This can be seen from the above divide-and-conquer method, where the introducing of the new variable r is the key point.

As an application of the CRT, a modular approach to the multiplication of large integers was developed by Schönhage [80]. Schönhage's algorithm has a complexity of $O(n^{1+(\sqrt{2}+\epsilon)/\sqrt{\log_2 n}})$ for the multiplication of two n-digit integers.

The idea of the Schönhage algorithm is a typical application of the modular technique depicted in Figure 3.1. Let x and y be two n-bit integers. Choose k pairwise relatively prime moduli $m_1, ..., m_k$ such that $m_i \geq 2$ and

$$m = \sum_{i=1}^{k} m_i \geq 2^{2n}. \tag{3.5}$$

The idea of the algorithm is described by the following steps:

Step 1: Compute for each $1 \leq i \leq k$

$$x_i = x \bmod m_i, \quad 0 \leq x_i \leq m_i - 1,$$
$$y_i = y \bmod m_i, \quad 0 \leq y_i \leq m_i - 1.$$

Step 2: Compute for each $1 \leq i \leq k$

$$z_i = x_i y_i \bmod m_i, \quad 0 \leq z_i \leq m_i - 1.$$

Step 3: Solve the following Chinese Remainder Problem of finding the z such that

$$z = z_i \bmod m_i, \quad 1 \leq i \leq k,$$

where $0 \leq z \leq m - 1$.

The purpose of (3.5) is to ensure that $xy < m$ so that $xy \bmod m = xy$, i.e., there is no overflow when multiplying two n-bit integers. Naturally, there is no overflow for the addition and subtraction of two n-bit integers. Thus, the arithmetic of $\mathbf{Z}/(m)$ for two n-bit integers is the same as that of $\mathbf{Z}$ for two n-bit integers. This is a necessary condition for the correctness of Step 3. The above three steps constitute only a general description of the Schönhage algorithm, details of the choices of the k, the moduli, and the computation of each step will be described in detail in what follows.

Within any modular approach one basic problem is the choice of those moduli. For two positive integers g and h it is easily verified that

$$\gcd(2^g - 1, 2^h - 1) = 2^{\gcd(g,h)} - 1.$$

One of the key ideas of the Schönhage algorithm is the choice of a set of special moduli m_i with

$$m_i = 2^{q_i} - 1 = \sum_{j=0}^{q_i-1} 2^j,$$

where $q_1, ..., q_k$ are pairwise relatively prime. Such a choice of the special moduli is to ensure that the arithmetic of $\mathbf{Z}/(m_i)$ can be easily carried out. Without loss of generality we assume that the k pairwise relatively prime integers are ordered as follows:

$$1 < q_1 < q_2 < \cdots < q_k.$$

To satisfy the condition of (3.5), it is enough to have

$$\sum_{i=1}^{k} q_i \geq 2n + 1. \tag{3.6}$$

To see this, we note $q_i \geq i + 1$. Then

$$1 - \frac{1}{2^{q_i}} \geq 1 - \frac{1}{2^{i+1}} > 2\frac{2^i - 1}{2^{i+1} - 1}$$

and

$$\prod_{i=1}^{k} \left(1 - \frac{1}{2^{q_i}}\right) > 2^k \frac{2 - 1}{2^{k+1} - 1} > \frac{1}{2}.$$

It follows that

$$m = \prod_{i=1}^{k} (2^{q_i} - 1) > 2^{\sum q_i - 1} > 2^{2n}.$$

The parameter k contributes to the complexity of the Schönhage algorithm, and its optimal choice will be specified later. The parameters q_i are chosen as follows:

$$q_1 = \left\lfloor \frac{2n}{k} \right\rfloor + 1,$$
$$q_{i+1} = \min\{q | q > q_i \text{ and } \gcd(q, q_j) = 1 \text{ for } j \leq i\}. \tag{3.7}$$

Thus, condition (3.6) is satisfied. It is quite intuitive that the q_i should be chosen as small as possible under the condition of (3.6) and the condition of pairwise relative primality. The optimal choice of the q_i's depends on the distribution of prime numbers. However, at least we can choose the first k primes which are greater than $2n/k$. It is known that the number of primes less than q_k can be estimated from above by the polynomial $c_1 q_k^{5/8}$, where c_1 is a constant, so we have

$$q_k \leq \frac{2n}{k} + kc_1 q_k^{5/8}.$$

Let

$$\alpha = \frac{q_k}{\left(\frac{2n}{k}\right)}.$$

Then it is proved in [80] that for k small enough

$$\alpha \leq \frac{1}{1 - c_2 \frac{k^{11/8}}{n^{3/8}}} \leq 1 + c_3 \frac{k^{11/8}}{n^{3/8}},$$

where c_2, c_3 are constants. Then we have an estimation for q_k:

$$q_k \leq \frac{2n}{k}\left(1 + \frac{1}{n^{1/8}}\right) \tag{3.8}$$

when

$$c_3 k^{11/8} \leq n^{1/4} \text{ and } n \geq n_0, \tag{3.9}$$

where n_0 is a constant. The restriction (3.9) is not important, since the integer k will be chosen such that $k = n^{o(1)}$.

The basic operations of the modular arithmetic are addition, subtraction, and multiplication modulo m_i. For a special modulus $m_j = 2^{q_j} - 1$, these operations become simpler. For such special moduli, Schönhage suggested to relax the condition $0 \leq x_j < m_j$ slightly by allowing x_j to take on m_j as an alternative to $x_j = 0$, since this does not affect the homomorphic property of the mapping φ in the CRT. Thus, the residue x_j of x modulo m_j is allowed to be any q_j-bit number. With this stipulation the addition modulo m_j and the multiplication modulo m_j, denoted here by $\oplus_j$ and $\otimes_j$ respectively, become the following [46]

$$x_j \oplus_j y_j = \begin{cases} x_j + y_j, & \text{if } x_j + y_j < 2^{q_j}; \\ ((x_j + y_j) \bmod 2^{q_j}) + 1, & \text{otherwise,} \end{cases} \tag{3.10}$$

$$x_j \otimes_j y_j = (x_j \times y_j \bmod 2^{q_j}) \oplus_j \left\lfloor \frac{x_j \times y_j}{2^{q_j}} \right\rfloor, \tag{3.11}$$

where $+$ and $\times$ are the usual addition and multiplication of integers. The modulo-2^q operation is very simple. However, the arithmetic of $\mathbf{Z}/(2^q - 1)$ relies on the multiplications of two q-bit integers.

Step 1 of the Schönhage algorithm is to get the modular representation of the integers x and y modulo $m_j = 2^{q_j} - 1$. Let $B = 2^{q_j}$. Then each integer x can be expressed as

$$x = b_t B^t + b_{t-1} B^{t-1} + \cdots + b_1 B + b_0, \tag{3.12}$$

where $0 \leq b_i < B$ for $0 \leq i \leq t$. This is the B-adic expansion of x. Since $B = 1 \bmod (2^{q_j} - 1)$, we have

$$x = b_t \oplus_j b_{t-1} \oplus_j \cdots b_1 \oplus_j b_0. \tag{3.13}$$

With this approach Step 1 reduces to a number of divisions.

Step 2 is the calculation of $x_i y_i \bmod m_i$. With the procedure of (3.11) the main computational work within this step is the computation of the multiplications $x_i y_i$. In fact, any procedure of calculating the value of $x_i y_i \bmod m_i$ must rely on the multiplications $x_i y_i$. If the moduli m_i are small enough, then the best algorithms for the multiplication $x_i y_i$ are still the classical algorithms like the school method. Thus, in the Schönhage algorithm a number n_1 is chosen. If for some m_i the corresponding q_i is larger than n_1, then the multiplication of two integers of $\lceil \log m_i \rceil = q_i$ bits is done by the same Schönhage method recursively until the multiplications needed involve only integers having no more than n_1 bits. The multiplication of two integers having no more than n_1 bits is performed by the school method. Schönhage suggested to choose $k \geq 32$ and to choose the q_i such that $q_k < 2q_1$. If the q_i chosen according to (3.7) do not satisfy this condition, the school method is used for multiplication.

Step 3 is the solving of the Chinese Remainder Problem (briefly, CRP). Schönhage uses the usual CRA to solve the CRP, but detailed computations within the CRA are carefully designed. Since m_i and m_j are pairwise relative prime for $i \neq j$, with Euclidean algorithm we can find out an integer $S_{i,j}$ such that

$$S_{i,j} m_j = 1 \bmod m_i, \quad z_i \prod_{j=1, j \neq i}^{k} S_{i,j} m_j = z_i \bmod m_i. \tag{3.14}$$

Then we calculate

$$w_i = z_i \prod_{j \neq i} S_{i,j} \bmod m_i, \quad 0 \leq w_i \leq m_i. \tag{3.15}$$

Furthermore,

$$w_i \prod_{j \neq i} m_j = \begin{cases} z_i \bmod m_i, \\ 0 \bmod m_j \text{ for } j \neq i. \end{cases}$$

It follows that

$$w = \sum_{i=1}^{k} w_i \prod_{j \neq i} m_j \tag{3.16}$$

is a solution of the CRP, i.e.,

$$w = z \bmod m.$$

Note that $w_i \leq m_i$ and we get $w \leq km$. Finally, by division we get z from w. This is the outline of the CRA.

Within the above CRA, there are a number of special computations since the moduli m_i are of special form. The first computational work is the calculation of the parameters $S_{i,j}$ satisfying (3.14). Applying the Euclidean algorithm directly to the moduli m_i is one way to compute the $S_{i,j}$. But a much better method due to Schönhage is available with the help of the following lemma whose proof is easy.

Lemma 3.2.1 *Let $p > 1, q > 1$, $\gcd(p, q) = 1$, $P = 2^p - 1$, $Q = 2^q - 1$. And let the integers r and s be defined by*

$$r = p \bmod q, \quad 1 \leq r < q;$$
$$sp = 1 \bmod q, \quad 1 \leq s < q.$$

Define

$$S = \sum_{i=0}^{s-1} 2^{ir}.$$

Then $\gcd(P, Q) = 1$ and $SP = 1 \bmod Q$.

By this lemma the parameters $S_{i,j}$ can be calculated as follows. Apply the Euclidean algorithm to the q_i to get the parameters $s_{i,j}$ and $r_{i,j}$ satisfying

$$r_{i,j} = q_j \bmod q_i, \quad 1 \leq r_{i,j} < q_i;$$
$$s_{i,j} q_j = 1 \bmod q_i, \quad 1 \leq s_{i,j} < q_i.$$

Then the parameters $S_{i,j}$ are given by

$$S_{i,j} = \sum_{v=0}^{s_{i,j}-1} 2^{vr_{i,j}} .$$

Since $m_i = 2^{q_i} - 1$, this method is much better than applying the Euclidean algorithm directly to the m_i. So the first computation within Step 3 is well done.

One main computational work within Step 3 is the calculation of $k(k-1)$ products

$$v = u \sum_{i=0}^{s-1} 2^{ir} \bmod (2^q - 1), \quad 0 \le u < 2^q, \quad 0 \le v < 2^q \qquad (3.17)$$

in (3.15). For the calculation Schönhage developed a special algorithm.

Let $g = \lceil \log_2 s \rceil$, and the binary representation of s be

$$s = \sum_{j=0}^{g} \tau_j 2^j ,$$

where $g \le \log_2 s + 1 \le \log_2 q$. It is not hard to verify the following equality

$$u \sum_{j=0}^{s-1} 2^{jr} = \sum_{j=0}^{g} \tau_j 2^{r \sum_{l=0}^{j-1} \tau_l 2^l} \left\{ u \sum_{h=0}^{2^j-1} 2^{hr} \right\} . \qquad (3.18)$$

The computation of the products in (3.17) can be carried out by introducing the auxiliary variables defined by

$$\begin{cases} u_j = u \sum_{i=0}^{2^j-1} 2^{ir} \bmod (2^q - 1), & 0 \le u_j < 2^q, \\ a_j = 2^j r \bmod q, & 0 \le a_j < q, \\ b_j = r \sum_{i=0}^{j-1} \tau_i 2^i \bmod q, & 0 \le b_j < q, \\ v_j = \sum_{i=0}^{j} \tau_i 2^{b_i} u_i \bmod (2^q - 1), & 0 \le v_j < 2^q. \end{cases} \qquad (3.19)$$

With the help of the following equality

$$2^{\lambda q + e} = 2^e \bmod (2^q - 1) \qquad (3.20)$$

we have the following iterative procedure for the calculation of the products in (3.17):

$$\begin{cases} a_0 = r, & a_{j+1} = 2a_j \bmod q, \\ u_0 = u, & u_{j+1} = u_j + 2^{a_j} u_j \bmod (2^q - 1), \\ b_0 = 0, & b_{j+1} = b_j + \tau_j a_j \bmod q, \\ v_0 = \tau_0 u_0, & v_j = v_{j-1} + \tau_j 2^{b_j} u_j \bmod (2^q - 1). \end{cases} \qquad (3.21)$$

It then follows from (3.17), (3.18) and (3.19) that

$$v_g = v.$$

Thus, the iterative procedure described in (3.21) computes the v in (3.17).

Another main calculation within Step 3 is the computation of w with the expression of (3.16). The brute-force method requires $k(k-1)$ multiplications with $m_j = 2^{q_j} - 1$. This number can be reduced if k is of some special form, especially $k = 2^N$. To this end, Schönhage developed a recursive procedure.

Let $k = 2^N$ and

$$W(\mu,\nu) = \sum_{i=\mu 2^\nu+1}^{(\mu+1)2^\nu} w_i \prod_{j=\mu 2^\nu+1, j\neq i}^{(\mu+1)2^\nu} (2^{q_j} - 1).$$

The recursive procedure is described by

$$\left\{ \begin{array}{ll} W(\mu,0) & = \; w_{\mu+1}, \\ W(\mu,\nu+1) & = \; W(2\mu,\nu) \prod_{j=(2\mu+1)2^\nu+1}^{(\mu+1)2^{\nu+1}} (2^{q_j}-1)+ \\ & + \; W(2\mu+1,\nu) \prod_{j=\mu 2^{\nu+1}+1}^{(2\mu+1)2^\nu} (2^{q_j}-1). \end{array} \right. \qquad (3.22)$$

The w we want to calculate is given by $W(0,N)$. It is not hard to see that the above recursive procedure needs only $kN = k\log_2 k$ multiplications for the calculation of the w.

For the case $k = 8$ the above recursive procedure can be illustrated by

$$\begin{aligned} w = & \; \{(w_1 m_2 + w_2 m_1)m_3 m_4 + (w_3 m_4 + w_4 m_3)m_1 m_2\}m_5 m_6 m_7 m_8+ \\ & + \{(w_5 m_6 + w_6 m_5)m_7 m_8 + (w_7 m_8 + w_8 m_7)m_5 m_6\}m_1 m_2 m_3 m_4. \end{aligned}$$

It is clear that only $24 = 8\log_2 8$ multiplications are needed.

So far the three steps in the Schönhage algorithm should be clear. It was shown in [80] that an optimal choice for k is $k = 2^N$ with

$$N = \left\lfloor \sqrt{2\log_2 n} \right\rfloor + 5.$$

Based on a model of Turing machines, the complexity of the Schönhage algorithm is of order $O(n^{1+(\sqrt{2}+\epsilon)/\sqrt{\log_2 n}})$. We refer to [80] for details of the complexity analysis, and to [76] for Turing machines and complexity theory.

3.3   Computing Exact Polynomial Resultants

Let R be a commutative ring with an identity and let A and B be polynomials over R. If

$$A(x) = \sum_{i=0}^{m} a_i x^i, \quad B(x) = \sum_{i=0}^{n} b_i x^i,$$

where $\deg(A) = m$ and $\deg(B) = n$, the *Sylvester matrix* of A and B is defined to be the $m + n$ by $m + n$ matrix

$$M = \begin{bmatrix}
a_m & a_{m-1} & \cdots & a_0 & 0 & 0 & 0 & 0 \\
0 & a_m & a_{m-1} & \cdots & a_0 & 0 & 0 & 0 \\
\vdots & \vdots & \vdots & \vdots & \vdots & \vdots & \vdots & \vdots \\
0 & 0 & 0 & a_m & a_{m-1} & a_{m-2} & \cdots & a_0 \\
b_n & b_{n-1} & \cdots & b_1 & b_0 & 0 & 0 & 0 \\
0 & b_n & b_{n-1} & \cdots & b_1 & b_0 & 0 & 0 \\
\vdots & \vdots & \vdots & \vdots & \vdots & \vdots & \vdots & \vdots \\
0 & 0 & 0 & 0 & b_n & b_{n-1} & \cdots & b_0
\end{bmatrix}$$

in which there are n rows of A coefficients, m rows of B coefficients. The *resultant* of A and B, denoted by $\operatorname{res}(A, B)$, is the determinant of the Sylvester matrix, which is an element of R. Collins [22] extended the definition to the case where either m or n, but not both, is zero. If, for example, $n = 0$, then M consists of just m rows of B coefficients.

Polynomials in one variable are called *univariate* and those in no less than two variables are referred to as *multivariate*. The resultant calculation has applications in systems of polynomial equations [48, 49, 63] and in quantifier elimination [44, pp.295-316].

If $R = \mathbf{Z}$, then the resultant $\operatorname{res}(A, B)$ of two polynomials over $\mathbf{Z}[x]$ involves only the addition, subtraction, and multiplication of $\mathbf{Z}$. Thus, by the analysis of Section 3.1 it is possible to develop a modular algorithm for the calculation of the resultant $\operatorname{res}(A, B)$. This was done by Collins [22].

There are some basic properties about the resultant. It follows immediately from the definition that

$$\operatorname{res}(A, B) = (-1)^{mn} \operatorname{res}(B, A).$$

The following theorem due to Collins describes another fundamental property of the resultant [22].

Theorem 3.3.1 *Let R be a commutative ring with identity and let A and B be polynomials of positive degree over R. Then there exist polynomials S and T over R such that*

$$AS + BT = \operatorname{res}(A, B),$$

where $\deg(S) < \deg(B)$ and $\deg(T) < \deg(A)$.

Proof: Let $m = \deg(A)$ and $n = \deg(B)$. For $1 \leq i < m + n$, multiplying the ith column of the Sylvester matrix M by x^{m+n-i} and adding to the last column give a new matrix M' whose determinant is equal to the determinant of M, and whose last column consists of the polynomials $x^{n-1}A(x), x^{n-2}A(x), ..., A(x)$, $x^{m-1}B(x), x^{m-2}B(x), ..., B(x)$. Expanding the determinant of M' with respect to its last column, we obtain an identity

$$A(x)S(x) + B(x)T(x) = \det(M') = \det(M) = \text{res}(A, B),$$

where the coefficients of S and T are cofactors of the last column of M', and hence of M, and therefore belong to R.   □

Let A and B be polynomials of positive degrees over an integral domain D. Then it follows from Theorem 3.3.1 that $\text{res}(A, B) = 0$ if and only if there exist nonzero polynomials S and T over D such that

$$AS + BT = 0,$$

where $\deg(S) < \deg(B)$ and $\deg(T) < \deg(A)$.

Assume that A and B are polynomials of positive degrees over an integral domain D. By generalizing the proof in [97, pp.83-84] or [44, pp.298-299], one can prove that $\text{res}(A, B) = 0$ if and only if A and B have a common divisor of positive degree.

Let R and R° be commutative rings with identities and let ϕ be any homomorphism of R into R°. Then ϕ induces a homomorphism of $R[x]$ into $R^\circ[x]$, which is also denoted by ϕ, defined by

$$\phi(\sum_{i=0}^{m} a_i x^i) = \sum_{i=0}^{m} \phi(a_i) x^i.$$

Applying the idea of the modular and homomorphic computing techniques formulated in Section 3.1 to the problem of the calculation of the resultant of two polynomials over R, we choose a number of homomorphisms ϕ from R to R° and compute the images $\phi(\text{res}(A, B))$ of the resultant $\text{res}(A, B)$. If the number of the homomorphisms is large enough and those homomorphisms are chosen properly, the images $\phi(\text{res}(A, B))$ will provide enough information for the resultant $\text{res}(A, B)$. Thus we will be able to compute the resultant $\text{res}(A, B)$.

For our resultant problem, the first thing we should solve is the relation between $\phi(\text{res}(A, B))$ and $\text{res}(\phi(A), \phi(B))$. This is settled by the following theorem due to Collins [22].

Theorem 3.3.2 *Let*

$$A(x) = \sum_{i=0}^{m} a_i x^i, \quad B(x) = \sum_{i=0}^{n} b_i x^i$$

be polynomials over R with $\deg(A) = m > 0$ and $\deg(B) = n > 0$. If $\deg(\phi(A)) = m > 0$ and $\deg(\phi(B)) = k$, where $0 \le k \le n$, then

$$\phi(\text{res}(A, B)) = \phi(a_m)^{n-k} \text{res}(\phi(A), \phi(B)).$$

Proof: Let M be the Sylvester matrix of A and B, M° the Sylvester matrix of $A^\circ = \phi(A)$ and $B^\circ = \phi(B)$. If $k = n$, then by definition $\phi(M) = M^\circ$ and

$$\phi(\text{res}(A, B)) = \phi(\det(M)) = \det(M^\circ) = \text{res}(A^\circ, B^\circ),$$

as asserted. If $k < n$, then it is easily seen that M° can be obtained from $\phi(M)$ by deleting its first $n - k$ rows and columns. Since the first $n - k$ columns of $\phi(M)$ contain $\phi(a_m)$ on the diagonal and are zero below the diagonal, we have

$$\begin{aligned}
\phi(\text{res}(A, B)) &= \phi(\det(M)) = \det(\phi(M)) \\
&= \phi(a_m)^{n-k} \det(M^\circ) \\
&= \phi(a_m)^{n-k} \text{res}(A^\circ, B^\circ).
\end{aligned}$$

This completes the proof.   □

This theorem shows that the computation of $\phi(\text{res}(A, B))$ is the same as that of $\text{res}(\phi(A), \phi(B))$ up to a constant $\phi(a_m)^{n-k}$. Thus, we have to solve the computation of the resultant in the image domains, which are chosen to be $\mathbf{Z}/(p)$ for a number of primes.

The Collins modular algorithm for the calculation of the resultant of multivariate polynomials needs a variant of the Garner algorithm [46, pp.253-254] which is a variant of the CRA. The inputs to the Garner algorithm for two moduli are the two odd moduli Q and p, and two integers B and a, such that $|B| < Q/2$ and $0 \le a < p$. The output is the unique C such that

$$C = B \bmod Q, \quad C = a \bmod p$$

and $C < pQ/2$. The calculations are as follows:

Step 1: $b = B \bmod p$.

Step 2: $q = Q \bmod p$.

Step 3: $d = (a - b)/q$ (here the arithmetic is in $\mathbf{Z}/(p)$).

Step 4: If $2d > p$, replace d by $d - p$.

Step 5: $C = dQ + B$.

It is noted that if $Q = 1$ and $B = 0$, then C is the unique integer such that $C = a \bmod p$ and $C < p/2$. The Garner algorithm for prime moduli $p_1, p_2, ..., p_k$, where $k \geq 3$, is an iterative application of the above algorithm for successive pairs of moduli $(1, p_1), (p_1, p_2), (p_1 p_2, p_3), ..., (p_1 \cdots p_{k-1}, p_k)$. Thus, in general the first modulus is much larger than the second, which is a prime.

By applying this algorithm to pairs of corresponding coefficients one obtains a CRA for polynomials. We will refer to this algorithm as the *Garner polynomial algorithm*. Given the moduli Q and p, a multivariate polynomial B with integer coefficients less than $Q/2$ in magnitude, and a multivariate polynomial A over $\mathbf{Z}/(p)$, The Garner polynomial algorithm computes the unique polynomial A over $\mathbf{Z}/(p)$ such that $C \bmod Q = B$ and $C \bmod p = A$, and the integer coefficients of C are less than $pQ/2$ in magnitude.

As seen in Section 3.2, the Schönhage modular algorithm for the multiplication of two large integers relies on an algorithm for the multiplication $x_i y_i$ of integers x_i and y_i in the range $0 \leq x_i \leq m_i$ and $0 \leq y_i \leq m_i$, where m_i are the moduli, that is, the computation of a function A is only reduced to that of $\phi(A)$ in the image ring or image field for a number of modular homomorphisms ϕ. This is true for every modular algorithm based on the CRT. In the Collins modular algorithm the calculation of the resultant of two polynomials A and B over $\mathbf{Z}$ is first reduced to that of the two polynomials $A \bmod p$ and $B \bmod p$ over $\mathbf{Z}/(p)$ for a number of prime p. Then the calculation of the resultant of two polynomials over $\mathbf{Z}/(p)$ is done by an algorithm which is called *CPRES* (*congruence polynomial resultant*) by Collins [22].

The algorithm CPRES is based on the *natural PRS resultant algorithm* or CPRES1 (congruence polynomial resultant, 1 variable) [22], which computes the resultant of two univariate polynomials over $\mathbf{Z}/(p)$. Let A and B be two univariate polynomials over $\mathbf{Z}/(p)$. The natural PRS (polynomial remainder sequence) generated by A and B is the sequence $A_1, A_2, ..., A_k$ defined by

$$A_1 = A, \quad A_2 = B,$$
$$A_i = Q_i A_{i+1} + A_{i+2}, \quad 1 \leq i \leq k - 2,$$

where $\deg(A_{i+2}) < \deg(A_{i+1})$, and Q_i is the quotient and A_{i+2} is the remainder, also $\deg(A_k) = 0$. Since $\mathbf{Z}/(p)$ is a field, such a PRS exists.

Let $c_i = \mathrm{ldcf}(A_i)$ and $n_i = \mathrm{ldcf}(Q_i)$ for $1 \leq i \leq k$, where $\mathrm{ldcf}(A_i)$ denotes the leading coefficient of A_i. It is proved in [15, 16] (see also [22]) that

$$\mathrm{res}(A, B) = (-1)^{\sum_{j=1}^{k-2} n_i n_{i+1}} \left[\prod_{i=2}^{k-1} c_i^{n_{i-1} - n_{i+1}} \right] c_k^{n_{k-1}}. \tag{3.23}$$

The CPRES1, which is also called the natural PRS resultant algorithm, is the computation of the natural PRS and the resultant with formula (3.23).

There is a unique homomorphism ϕ_m of $\mathbf{Z}$ onto $\mathbf{Z}/(m)$ such that

$$\phi_m(i) = i \ \text{ for } 0 \le i \le m,$$

which is called a *modular homomorphism*. It is given by $\phi_m(x) = x \bmod m$ for each $x \in \mathbf{Z}$. For any $r \ge 1$, ϕ_m induces a unique homomorphism ϕ'_m of $\mathbf{Z}[x_1, ..., x_r]$ onto $\mathbf{Z}/(m)[x_1, ..., x_r]$ such that

$$\phi'_m(a) = \phi_m(a) \ \text{ for all } a \in \mathbf{Z}$$

and

$$\phi'_m(x_i) = x_i \ \text{ for } 0 \le i \le r.$$

The mapping ϕ'_m is also called a modular homomorphism and denoted by ϕ_m.

Let R be the polynomial ring $R^\circ[x]$. Then each element of R° gives a homomorphism

$$\psi_a(A) = A(a), \ \ A(x) \in R = R^\circ[x].$$

This ψ_a is called an *evaluation homomorphism*. Generally, if R is a polynomial ring $R^\circ[x_1, ..., x_r]$ of polynomials in r variables, then by induction it is easy to see that the mapping

$$\psi_{(a_1, ..., a_r)}(A) = A(a_1, ..., a_r), \ \ A(x_1, ..., x_r) \in R^\circ[x_1, ..., x_r],$$

is a homomorphism, because of the natural isomorphism between $R^\circ[x_1, ..., x_r]$ and $(R^\circ[x_1, ..., x_{r-1}])[x_r]$. For any r-tuple $(a_1, ..., a_r)$ of elements of R° this mapping is also called an evaluation homomorphism.

Basing on the above CPRES1 Collins developed the algorithm CPRES which can compute the resultant of two polynomials over $\mathbf{Z}/(p)[x_1, ..., x_r]$, where $r \ge 2$. Let

$$R^\circ = \mathbf{Z}/(p)[x_1, ..., x_{r-2}]$$

(if $r = 2$, then we set $R^\circ = \mathbf{Z}/(p)$ and $R = R^\circ[x_{r-1}]$), and ψ_a be an evaluation homomorphism defined by any $a \in R^\circ$. Then

$$\phi_a(A) = A(x_1, ..., x_{r-2}, a, x_r)$$

and

$$\phi_a(B) = B(x_1, ..., x_{r-2}, a, x_r)$$

are polynomials over $\mathbf{Z}/(p)$ in r-variables. Let

$$A(x_1, ..., x_{r-2}, a, x_r) = \sum_{i=0}^{m} A_i(x_1, ..., x_{r-2}, a)x_r^i,$$

$$B(x_1, ..., x_{r-2}, a, x_r) = \sum_{i=0}^{n} B_i(x_1, ..., x_{r-2}, a)x_r^i.$$

By Theorem 3.3.2 if

$$A_m(x_1, ..., x_{r-2}, a) \neq 0,$$
$$B_m(x_1, ..., x_{r-2}, a) \neq 0,$$
$$C = \mathrm{res}(A, B),$$

then it follows that

$$\mathrm{res}(\psi_a(A), \psi_a(B)) = C(x_1, ..., x_{r-2}, a).$$

Hence, if we have a bound k on the degree of C in x_{r-1}, then we can compute C by interpolation from the resultants $\mathrm{res}(\psi_a(A), \psi_a(B))$ for $k+1$ different values of a. If $r > 2$ then R° is infinite and sufficiently many evaluation homomorphisms ψ_a satisfying the hypotheses of Theorem 3.3.2 will always be available. However, if $r = 2$ then R° has only p distinct elements a and this method will not be applicable if there are not enough primes. It is clear that even when $r > 2$ it is much more convenient and computationally superior, to use only ψ_a for which $a \in \mathbf{Z}/(p)$. Thus, in practice the prime p should be large enough.

With the above algorithm CPRES1 for the calculation of the resultant of two univariate polynomials over $\mathbf{Z}/(p)$, the algorithm CPRES for that of two multivariate polynomials over $\mathbf{Z}/(p)$ can be derived as follows. Let $A(x_1, ..., x_r)$ and $B(x_1, ..., x_r)$ be polynomials over $\mathbf{Z}/(p)$, where $r \geq 2$, and let $C(x_1, ..., x_{r-1})$ be their resultant with respect to x_r. If A and B have degree m_r and n_r in x_r, then the determinant C is the sum of at most $(m_r+n_r)!$ terms, each of which is a product of n_r A coefficients and m_r B coefficients. Hence if A and B have degree of m_r and n_r in x_r, then the degree of C in x_{r-1} is at most $m_r n_{r-1} + n_r m_{r-1}$. This bound is used in the following evaluation algorithm CPRES where polynomial interpolations are needed.

As seen above, an interpolation algorithm is needed to compute the $\mathrm{res}(A, B)$ for polynomials over $\mathbf{Z}/(p)$ in r variables, where $r \geq 2$. The interpolation algorithm employed in the Collins algorithm is an iterative method, analogous to the Garner polynomial algorithm above, which is a variant of the CRA. The inputs to this algorithm are a polynomial

$$D(x_r) = \sum_{i=0}^{k}(x_r - b_i),$$

where the b_i are distinct elements of $\mathbf{Z}/(p)$, an element b of $\mathbf{Z}/(p)$ distinct from b_i, a polynomial $A(x_1, ..., x_r)$ over $\mathbf{Z}/(p)$ of degree k or less in x_r, and a polynomial $C(x_1, ..., x_{r-1})$ over $\mathbf{Z}/(p)$ (if $r = 1$ then $C \in \mathbf{Z}/(p)$). The output is the unique polynomial $G(x_1, ..., x_r)$, of degree $k + 1$ or less in x_r, such that

$$G(x_1, ..., x_{r-1}, b_i) = A(x_1, ..., x_{r-1}, b_i)$$

for $0 < i \leq k$ and

$$G(x_1, ..., x_{r-1}, b) = C(x_1, ..., x_{r-1})$$

The interpolation algorithm computes the polynomial G by the formula

$$\begin{aligned}G(x_1, ..., x_r) &= \{C(x_1, ..., x_{r-1}) - A(x_1, ..., x_{r-1}, b)\}D(b)^{-1}D(x_r) \\ &+ A(x_1, ..., x_r).\end{aligned} \quad (3.24)$$

The special case $k = -1, D(x_r) = 1$ and $A(x_1, ..., x_r) = 0$ is permitted, in which case

$$G(x_1, ..., x_r) = C(x_1, ..., x_{r-1}).$$

Based on the above analysis, the algorithm CPRES due to Collins for the calculation of the resultant of two multivariate polynomials $A(x_1, ..., x_r)$ and $B(x_1, ..., x_r)$ over $\mathbf{Z}/(p)$ with respect to x_r can be described as follows.

Algorithm CPRES:

Inputs: Polynomials $A(x_1, ..., x_r)$ and $B(x_1, ..., x_r)$ over $\mathbf{Z}/(p)$ of positive degree in x_r.

Output: $C(x_1, ..., x_{r-1})$, the resultant of A and B with respect to x_r ($C \in \mathbf{Z}/(p)$ if $r = 1$).

Step 1: If $r = 1$, apply the algorithm CPRES1 above.

Step 2: Set

$$\begin{aligned}m_r &:= \deg(A) \text{ in } x_r; \\ n_r &:= \deg(B) \text{ in } x_r; \\ m_{r-1} &:= \deg(A) \text{ in } x_{r-1}; \\ n_{r-1} &:= \deg(B) \text{ in } x_{r-1}; \\ k &:= m_r n_{r-1} + n_r m_{r-1} + 1; \\ C(x_1, ..., x_{r-1}) &:= 0; \\ D(x_{r-1}) &:= 1, \\ b &:= -1.\end{aligned}$$

Step 3: Set $b := b + 1$; if $b = p$, stop and report failure.

Step 4: Set

$$A^\circ(x_1, ..., x_{r-2}, x_r) := A(x_1, ..., x_{r-2}, b, x_r);$$
$$B^\circ(x_1, ..., x_{r-2}, x_r) := B(x_1, ..., x_{r-2}, b, x_r);$$

if degree of A° in x_r is less than m_r or degree of B° in x_r is less than n_r, go to Step 3.

Step 5: Set $C^\circ = \text{res}(A^\circ, B^\circ)$. (The algorithm here applies itself recursively.)

Step 6: Apply the interpolation algorithm above to $D(x_{r-1}), b, C(x_1, ..., x_{r-1})$ and $C^\circ(x_1, ..., x_{r-2})$, obtaining $G(x_1, ..., x_{r-1})$.

Step 7: Set $C := G$ and replace $D(x_{r-1})$ by $(x_{r-1} - b) \cdot D(x_{r-1})$, if $\deg(D) \leq k$, go to Step 3; otherwise exit.

It should be noted that the algorithm applies itself recursively. This is also the case with the Schönhage algorithm. Generally, many modular algorithms apply themselves recursively. This is because the modular approach is only a reduction method. Many divide-and-conquer techniques are similar to modular techniques in the sense that problems with large input size are reduced to a number of problems with smaller size. Thus, many divide-and-conquer methods can apply themselves recursively.

With the algorithm CPRES we are ready to formulate the Collins modular algorithm for the calculation of the resultant of two multivariate polynomials over $\mathbf{Z}$.

Let ϕ_{m_i} denote the modular homomorphism defined by

$$\phi_{m_i}(x) = x \bmod m_i, \quad x \in \mathbf{Z}.$$

By the CRT if $\phi_{m_i}(\text{res}(A, B))$ is known for sufficiently many pairwise relatively prime moduli m_i, then $\text{res}(A, B)$ can be computed. Thus, by Theorem 3.3.2 it suffices to compute $\text{res}(\phi_{m_i}(A), \phi_{m_i}(B))$ for sufficiently many m_i for which $\phi_{m_i}(\text{ldcf}(A))$ and $\phi_{m_i}(\text{ldcf}(B))$ are nonzero. The number of moduli needed depends on a bound for the coefficients of the resultant, which will be specified soon. In what follows we assume that each modulus is a prime.

To derive the bound for the coefficients of the resultant, we need to define the *norm* of multivariate polynomials A over $\mathbf{Z}$, which is denoted by $\text{norm}(A)$. The norm of any integer a is defined to be the absolute value, i.e., $\text{norm}(a) = |a|$. If

$$A(x) = \sum_{i=0}^{m} a_i x^i,$$

then the norm of $A(x)$ is defined by

$$\text{norm}(A) = \sum_{i=0}^{m} \text{norm}(a_i).$$

By induction on the number of variables this defines $\text{norm}(A)$ for all nonzero A. The norm of zero is defined to be zero. The following basic properties about this norm function are easily verified:

1. $\text{norm}(A + B) \leq \text{norm}(A) + \text{norm}(B)$;

2. $\text{norm}(AB) \leq \text{norm}(A)\,\text{norm}(B)$;

3. $|a| \leq \text{norm}(A)$ if a is any numerical coefficient of A.

Now let

$$A(x_1, ..., x_r) = \sum_{i=0}^{m} A_i(x_1, ..., x_{r-1})x_r^i,$$

$$B(x_1, ..., x_r) = \sum_{i=0}^{n} B_i(x_1, ..., x_{r-1})x_r^i$$

and let $C(x_1, ..., x_{r-1})$ be the resultant of A and B with respect to x_r. Set

$$d = \max_{0 \leq i \leq m} \text{norm}(A_i) \text{ and } e = \max_{0 \leq i \leq n} \text{norm}(B_i).$$

Each nonzero term of the determinant of the Sylvester matrix, as the product of n A coefficients and m B coefficients, has a norm at most $d^n e^m$. Since there are at most $(m + n)!$ such terms, we have

$$\text{norm}(C) \leq (m + n)!d^n e^m,$$

and hence if a is any integer coefficient of C, then

$$|a| \leq (m + n)!d^n e^m.$$

In the Collins modular algorithm the Garner polynomial algorithm is used. In applying this algorithm only odd prime moduli $p_1, ..., p_k$ are used. Recall that the Garner algorithm computes the integer solution a satisfying

$$|a| < p_1 \cdots p_k/2.$$

Thus, to obtain the coefficients of C by the Garner algorithm it suffices to compute $\phi_{m_i}(C)$ for $p_1, ..., p_k$ such that

$$(m + n)!d^n e^m < p_1 \cdots p_k/2,$$

which is equivalent to

$$p_1 \cdots p_k > 2(m+n)! d^n e^m.$$

Here it is clear that if one wishes to choose k small, then some of the primes must be large enough; and similarly, if one wishes to use small moduli, then k must be large enough. In the Collins modular algorithm described below it is assumed that there is a precomputed list of sufficiently many odd primes so that in practice the list will not be exhausted.

The following is the complete Collins modular algorithm for the computation of the resultant of two multivariate polynomials over **Z**. Following Collins [22] we call it PRES (polynomial resultant).

Collins modular algorithm: PRES

Inputs: Polynomials

$$A(x_1, ..., x_r) = \sum_{i=0}^{m} A_i(x_1, ..., x_{r-1}) x_r^i,$$

$$B(x_1, ..., x_r) = \sum_{i=0}^{n} B_i(x_1, ..., x_{r-1}) x_r^i$$

with integer coefficients, $r \geq 1$, $A_m \neq 0$, $B_n \neq 0$, $m > 0, n > 0$, and a list L of distinct odd primes.

Output: $C(x_1, ..., x_{r-1})$, the resultant of A and B with respect to x_r (C is an integer if $r = 1$).

Step 1: Set

$$
\begin{aligned}
&m := \deg(A) \text{ in } x_r; \\
&n := \deg(B) \text{ in } x_r; \\
&d := \max_{0 \leq i \leq m} \text{norm}(A_i); \\
&e := \max_{0 \leq i \leq n} \text{norm}(B_i); \\
&Q := 1; \\
&C := 0; \\
&f := 2(m+n)! d^n e^m.
\end{aligned}
$$

Step 2: Let p be the next prime on the list L (if no more primes remain, stop and report failure).

Step 3: Compute

$$A^\circ := \phi_p(A);$$
$$B^\circ := \phi_p(B);$$

if degree of A° in x_r is less than m or degree of B° in x_r is less than n, go to Step 2.

Step 4: Use algorithm CPRES to compute $C^\circ := \text{res}(A^\circ, B^\circ)$.

Step 5: Use the Garner polynomial algorithm to compute the unique polynomial D with integer coefficients less than $pQ/2$ in magnitude such that $D = C \bmod Q$ and $D = C^\circ \bmod p$.

Step 6: Set $C := D$ and replace Q by pQ; if $Q \leq f$, go to Step 2; otherwise exit.

A multivariate polynomial $A(x_1, ..., x_r)$ over $\mathbf{Z}$ has a degree with respect to each x_i. Let l be the maximum degree of two multivariate polynomials in r variables. It is shown that the computational complexity of the algorithm PRES is $O((l+1)^{2r+1}(\log c(l+1)) + (l+1)^{2r}(\log c(l+1))^2$. We refer to [22] for the detailed complexity analysis.

3.4   Other Applications in Symbolic Computations

Symbolic computation is an important area of computer science. It is quite different from numerical computation, where high-precision approximations are allowable. In symbolic computation exact results are required at each step. One problem regarding some symbolic computations is that intermediate results are too large to be manipulated.

One example is the computation of the greatest common divisor (briefly, GCD) of polynomials. A specific example [46] is the computation of the GCD of the polynomials

$$F_1(x) = x^8 + x^6 - 3x^4 - 3x^3 + 8x^2 + 2x - 5,$$
$$F_2(x) = 3x^6 + 5x^4 - 4x^2 - 9x + 21,$$

where they are viewed as polynomials with rational coefficients. With the Euclidean algorithm we have the following results of computation:

$$x^8 + x^6 - 3x^4 - 3x^3 + 8x^2 + 2x - 5,$$
$$3x^6 + 5x^4 - 4x^2 - 9x + 21,$$

$$-\frac{5}{9}x^4 + \frac{1}{9}x^2 - \frac{1}{3},$$

$$-\frac{117}{25}x^2 - \frac{1}{9}x + \frac{441}{25},$$

$$\frac{1288744821}{543589225}.$$

It follows that F_1 and F_2 are relatively prime. The intermediate results of the computation are quite large, compared with the coefficients of F_1 and F_2. This phenomenon is referred to as the *coefficient growth*.

One algorithm which effectively controls the coefficient growth, without any GCD computations in the coefficient domain or any subdomain thereof, is the subresultant PRS algorithm due to Collins [21]. Another is the modular algorithm based on the CRA developed by Brown [17], which is markedly superior to the classical algorithm. With the general idea of modular techniques based on the CRA, it is not hard to derive a modular algorithm for the calculation of the GCD of multivariate polynomials of $\mathbf{Z}[x_1, ..., x_r]$. According to [22], the modular algorithm is markedly superior to the classical one.

The computation of the exact solutions of systems of linear equations with integral or rational or polynomial coefficients is important for many applications. Direct methods such as the Gaussian or Gauss-Jordan elimination [69] employ rational operations and are, in general, inefficient for exact computation. Eliminating variables by cross-multiplying leads to a tremendous growth in the size of intermediate results. The intermediate expression swell is well illustrated in an example of Rosser [73], in which a 6×6 matrix is given whose elements are integers in the interval $[-192, 195]$. Intermediate results of the order of 10^{29} are obtained after cross-multiplying.

To avoid having intermediate expression swell, modular algorithms for computing the exact solution of systems of linear equations have been developed by Borrosh and Frankel [13] and by McClellan [59]. Those modular algorithms are superior to classical methods. With the general idea of modular techniques described in Section 3.1 it is not hard to develop a modular algorithm or to derive the known algorithms for the computation of the exact solutions of systems of linear equations with integral or polynomial coefficients, though there are some minor specific details to be taken care of.

3.5 CRA and Homomorphic Image Computing

As seen in the foregoing sections, modular techniques are a generalization and extension of the CRA. The key idea is the conversion of the original computation into the computation of the homomorphic images under a number of special modular homomorphisms

$$\phi_m(y) = y \bmod m, \quad y \in \mathbf{Z} \text{ or } \mathbf{Z}[x_1, ..., x_r].$$

This idea can be extended into a kind of more general computing techniques, which is called the *homomorphic image computing*. In this section the homomorphic image computing and its problems will be outlined.

We will formulate the ideas of homomorphic image computing in terms of algebraic systems (*universal algebras*), also referred to as Ω-algebras. An algebraic system *(algebra, Ω-algebra)* is a pair $[A; \Omega]$ where

1. A is a set (the "carrier" of the algebraic system).

2. Ω is a set of operations defined on A: each operation $\omega \in \Omega$ is a function $A^n \to A$.

If Ω is clear, we denote $[A; \Omega]$ simply by A. Important examples of Ω-algebras are monoids $[M; *, 1]$, semigroups, groups $[G; \times, 1]$, rings (including integral domain and Euclidean domain) and fields.

Let A be an Ω-algebra. A subset B of A is called a (Ω-) *subalgebra* of A if B is Ω-closed: for any operation $\omega \in \Omega$,

$$x_1, ..., x_n \in B \Rightarrow w(x_1, ..., x_n) \in B.$$

Examples are subgroups, subsemigroups, subrings. Let A be an Ω-algebra and H a subset of A. The smallest subalgebra that contains H is called the *subalgebra generated by H*, and denoted by $[H]$.

We call two algebras $[A; \Omega]$, $[A'; \Omega']$ *similar* if there is a bijection from Ω to Ω' with the property that corresponding operations $\omega \in \Omega$, $\omega' \in \Omega'$ have the same arity. Let $[A; \Omega]$ and $[A'; \Omega']$ be similar algebras. A function $\phi : A \to A'$ is said to be a *morphism* (also called *homomorphism*) if for any $\omega \in \Omega$ and for any $a_1, ..., a_n \in A$,

$$\phi(\omega(a_1, ..., a_n)) = \omega'(\phi(a_1), .., \phi(a_n)).$$

Note that homomorphisms for Ω-algebras are a generalization of group, ring and field homomorphisms.

A quotient set A/E of an Ω-algebra A by a special kind of equivalence relation is called a *congruence relation*, with operations defined on A/E in a natural way. A *congruence relation E* on an Ω-algebra A is an equivalence relation on A that satisfies the following *substitution property*: for any $\omega \in \Omega$,

$$a_i E b_i \text{ for } i = 1, ..., n \Rightarrow \omega(a_1, ..., a_n) E \omega(b_1, .., b_n).$$

Let A be an Ω-algebra and E a congruence relation on A. Just as for groups and rings, it is easy to prove

(a) The quotient set A/E is an Ω-algebra if we define $\omega \in \Omega$ on A/E by

$$\omega([a_1], ..., [a_n]) = [\omega(a_1, ..., a_n)];$$

(b) A/E is a homomorphic image of A under the natural map $\nu : a \mapsto [a]$.

Let $\phi : A \to A'$ be a morphism of Ω-algebras. Then the kernel relation E_ϕ,

$$aE_\phi b \Leftrightarrow \phi(a) = \phi(b),$$

is a congruence relation on A.

Assume that A is an Ω-algebra, and $F(x_1, ..., x_t)$ an algebraic expression over A, by which we mean that it involves only operations of Ω, and $x_1, ..., x_t$ are variable inputs. To compute the value F of this algebraic expression, the idea of a homomorphic image computing technique may be described as follows.

Step 0: Find a number of Ω_i-algebras A_i and homomorphisms ϕ_i from A to A_i for $i = 1, ..., k$.

Step 1: For $i = 1$ to k compute $\phi_i(x_1), ..., \phi_i(x_t)$ and the homomorphic algebraic expressions F_i over Ω_i of F under ϕ_i. The algebraic expression F_i over Ω_i is obtained by applying ϕ_i to each operation and each coefficient in the algebraic expression of F.

Step 2: For $i = 1$ to k compute $F_i := F_i(\phi_i(x_1), ..., \phi_i(x_t))$.

Step 3: Use a method to compute F from the images F_i.

It is clear that the idea of homomorphic image computing is a generalization of that of modular techniques based on the CRA.

The first problem of the homomorphic image computing is the choice of the Ω_i-algebras A_i and the homomorphisms ϕ_i. In modular techniques based on the CRA, the Ω_i-algebra A_i are chosen to be the rings $\mathbf{Z}/(m_i)$ or $\mathbf{Z}/(m_i)[x_1, ..., x_r]$ for example. There are some general principal for the choices. They should be chosen such that

- the computation of an algebraic expression over A_i should be easier than that of all of its preimages over A with respect to the homomorphism ϕ_i;

- the images F_i should provide as much information about F as possible; and

- an algorithm for computing F from its images F_i should be available.

In the modular computing the operations of $\mathbf{Z}/(m_i)$ (resp. $\mathbf{Z}/(m_i)[x_1, ..., x_r]$) are usually easier than those of $\mathbf{Z}$ (resp. $\mathbf{Z}[x_1, ..., x_r]$). In addition, the moduli are chosen to be pairwise relatively prime so that the images provide the maximum possible amount of information about their preimage. Also the modular homomorphisms make it possible to apply the CRA or one of its variants to compute the value of an arithmetic expression from its images.

The second problem of the homomorphic image computing is the development of an algorithm for the computation of F from its images F_i. Such an algorithm should have a reasonable complexity.

The evaluation of a specific homomorphic image computing technique should be based on the computational complexity of the whole homomorphic image computing algorithm. The homomorphic image computing constitutes only a general framework, where specific details have to be filled in.

3.6   Information and CRT

In a modular algorithm the moduli from $\mathbf{Z}$ or $F[x]$ are usually required to be pairwise relatively prime. We have made such a requirement throughout the whole chapter without making the reason clear. Before explaining this requirement, we shall in this section have a look at the CRT from the information point of view.

Consider the arithmetic expression $A(x, y, x) = x + yz \in \mathbf{Z}[x, y, z]$. Suppose that we are asked to compute $A(10, 9, 12)$. If we take the modulus 27 and consider the homomorphic image of A with respect to this modulus, we have $a_1 = A \bmod 27 = 10$. Note that $0 \leq A(10, 9, 12) < A(12, 12, 12) = 156$. Under the prior knowledge $0 \leq A(10, 9, 12) \leq 156$ the uncertainty of A is $\log_2 157$ bits. Thus, the modular image $a_1 = 10$ of $A(10, 9, 12)$ and the knowledge $0 \leq A(10, 9, 12) \leq 156$ give us the information that the value $A = A(10, 9, 12)$ is one of the elements of the set of 18 elements

$$S_1 = \{10, 37, 64, 91, 118, 145\}.$$

Therefore, the image a_1 reduces the uncertainty of A to $\log_2 6$ bits, and provides $\log_2 157 - \log_2 6 = \log_2 157/6$ bits of information about A.

Then we choose another modulus 9 and consider the homomorphic image $a_2 = A \bmod 9 = [(10 \bmod 9) + (9 \bmod 9)(12 \bmod 9)] \bmod 9 = 1$. This second homomorphic image gives us the information that the value $A = A(10, 9, 12)$ is one of the elements of the following set

$$S_2 = \{1, 10, 19, 28, 37, 46, 55, 64, 73, 82, 91, 100, 109, 118, 127, 136, 145, 154\}.$$

With the homomorphic image $a_2 = 1$ with respect to the modulus 9 the uncertainty of A is $\log_2 18$ bits. Thus, the amount of information about A that the homomorphic image a_2 provides is

$$\log_2 157 - \log_2 18 = \log_2 157/18 \text{ bits.}$$

Hence the homomorphic image with respect to 9 does provide information about A, but it does not provide further information about A if we know a_1. This is clear since $S_1 \subset S_2$. Behind this phenomenon is the fact that 9 divides 27. Generally, we have the following conclusion whose proof has already been implied in the above analysis.

Theorem 3.6.1 *Let $A(x_1, ..., x_t)$ be an arithmetic expression over the ring $\mathbf{Z}$ (resp. $GF(q)[x]$), and let m_1 and m_2 be two moduli from $\mathbf{Z}$ (resp. $GF(q)[x]$). Define for $i = 1, 2$*

$$a_i = A(x_1, ..., x_t) \bmod m_i.$$

If m_2 divides m_1, then $I(A; (a_1, a_2)) = I(A; a_1)$, where $I(A; B)$ denotes the amount of mutual information between A and B; i.e., the homomorphic image a_2 provides no further information about A if a_1 is known.

For the efficiency of a modular algorithm, the homomorphic images with respect to a set of moduli should provide as much information about their preimage as possible under the condition that the product of the moduli is fixed. Of course, the larger a modulus, the more information about the original arithmetic expression the homomorphic image of this arithmetic expression with respect to the modulus provides. Suppose that we are allowed to choose two moduli m_1, m_2 such that $m_1 m_2 = 36, m_1 \neq m_2$ and $m_i > 1$. Then we have three possibilities

$$\{m_1, m_2\} = \{4, 9\}, \ \{2, 18\}, \ \text{or} \ \{3, 12\}.$$

The first choice $\{m_1, m_2\} = \{4, 9\}$ is the best, since $\gcd(4, 9) = 1$ is the smallest among the greatest common divisors, and the last one is the worst since $\gcd(3, 12)$ is the largest. Generally, we have the following conclusion.

Theorem 3.6.2 *Let $m_1, m_2, ..., m_t$ be t positive integers with $m_i > 1$ for $i = 1, 2, ..., t$, and $l = \text{lcm}\{m_1, ..., m_t\}$. Also let B_1 and B_2 be two integers with $B_1 < B_2$. Under the prior knowledge $B_1 \leq A \leq B_2$ the t homomorphic images $a_i = A \bmod m_i$ provide*

$$\log_2(B_2 - B_1 + 1) - \log_2 \left(\left\lfloor \frac{B_2 - u}{l} \right\rfloor - \left\lceil \frac{B_1 - u}{l} \right\rceil + 1 \right)$$

$$= \log_2 \frac{B_2 - B_1 + 1}{\left\lfloor \frac{B_2 - u}{l} \right\rfloor - \left\lceil \frac{B_1 - u}{l} \right\rceil + 1} \text{ bits}$$

of information about A, where $0 \leq u < l$ is the unique solution of

$$u = x_i \bmod m_i \ \ for \ i = 1, 2, ..., t. \tag{3.25}$$

Proof: By the CRT we have

$$\mathbf{Z}/(l) \cong \mathbf{Z}/(m_1) \times \cdots \times \mathbf{Z}/(m_t).$$

Thus, for each set of t integers x_i with $0 \leq x_i < m_i$ there is a unique integer u in the range $0 \leq u < l$ which is a solution of the set of equations given by (3.25). In fact this u can be calculated by the extended CRA given in Section 2.4. It follows from the CRT that $A = lh + u$ for some integer h in the range $\lfloor (B_1 - u)/l \rfloor \leq h \leq \lceil (B_2 - u)/l \rceil$. Then the conclusion of the theorem follows. $\square$

By this theorem the larger the least common multiple l of the t moduli is, the more information about A the t homomorphic images a_i provide. Since $\mathrm{lcm}\{m_1, ..., m_t\} \leq \prod_{i=1}^{t} m_i$ and the equality holds if and only if these m_i are pairwise relatively prime, it is clear why the moduli for a modular algorithm should be chosen to be pairwise relatively prime.

The above discussion about CRT and information can be extended to other homomorphisms. To illustrate this, we consider an example. Consider some homomorphisms from the linear space $GF(q)^n$ to $GF(q)$, of which the former has dimension n and the latter has dimension one. Let

$$\phi_i(\mathbf{x}) = \mathbf{x}\mathbf{v}_i^T = \sum_{j=1}^{n} b_j v_{ij}, \ \ i = 1, 2, \cdots, m,$$

where $\mathbf{v}_i = (v_{1i}, \cdots, v_{ni})$ are nonzero vectors of $GF(q)^n$ and $\mathbf{x} = (x_1, \cdots, x_n) \in GF(q)^n$. It is obvious that each ϕ_i is a homomorphism. Suppose that the variable $\mathbf{x}$ takes on each element of $GF(q)^n$ equally likely. Then the uncertainty of $\mathbf{x}$ is $\log q^n$. Given the image $\phi_i(\mathbf{x})$, the uncertainty of $\mathbf{x}$ becomes $\log q^{n-1}$ since the number of solutions of $\phi_i(\mathbf{x}) = y$ is q^{n-1}. Thus, the amount of information about $\mathbf{x}$ the homomorphic image $\phi_i(\mathbf{x})$ provides is $\log q$. It follows that the homomorphic image of each such homomorphism provides the same amount of information about the preimage. It is easy to see that the homomorphic images $\phi_1(\mathbf{x}), \cdots, \phi_m(\mathbf{x})$ determine $\mathbf{x}$ if and only if the rank of the vectors $\mathbf{v}_i$ is n.

Chapter 4

In Algorithmics

Chinese Remainder Algorithm is not only itself a divide-and-conquer technique, but can also be used to develop other divide-and-conquer techniques for solving other complicated problems. This can be illustrated by the linear-feedback shift-register synthesis problem for sequences over $\mathbf{Z}/(m)$ outlined in Section 4.4. Also interesting are variants of the CRA which solve some problems directly. One of such examples is the polynomial interpolation problem over fields described in Section 4.2. One main topic of this chapter is the application of CRA and CRT in developing such algorithms.

Cyclic convolution and the discrete Fourier transform play an important role not only in digital signal processing, but also in mathematics and computer science in general. There are currently quite a number of algorithms for the calculation of the cyclic convolution and of the discrete Fourier transform. It seems that most of those algorithms are based on the idea of rearranging one-dimensional arrays into multidimensional arrays with shorter lengths. There are a number of ways of doing so. One of the most effective ways to rearrange a one-dimensional array into a multidimensional one is based on the CRT and CRA.

Another main topic of this chapter is the data rearrangement technique based on the CRT and its applications in developing algorithms for cyclic convolution and the discrete Fourier transform. It is interesting to note that the cyclic convolution and the discrete Fourier transform involve only the operations of some fields, whereas the indices of those data arrays are integers. Thus, it is very often the case that the CRT and CRA cannot be used to manipulate those data arrays over general fields, but they can certainly be employed to manipulate the integral indices of those data arrays. This could result in fast algorithms in certain cases. While the discrete Fourier transform (DFT) has wide applications, the CRT transform, including the DFT as a special case, could be more important.

4.1   Divide-and-Conquer Techniques

Divide-and-conquer techniques, which are often employed in our daily life, are quite useful in problem solving. There are two main steps within a divide-and-conquer technique. The first step is the dividing of a problem into subproblems. The second is the solving of those subproblems and the synthesizing of the solutions of subproblems to get solutions of the original problem.

The dividing step of a divide-and-conquer technique is crucial. To illustrate this, let us take an example. Suppose we are told that there is an integer x between 1 and 32 which is in Alice's pocket, and we are asked to guess the actual value of x with the minimum number of yes-no questions to Alice. Each time we are allowed to ask Alice any question which can only be answered by Alice with yes or no. One naive divide-and-conquer scheme is to ask Alice whether the value of x is 1, 2, ..., 32, in turn. The worst-case complexity of this divide-and-conquer scheme is 32, this means that we have to ask Alice 32 questions in the worst case. There are divide-and-conquer techniques with which we can get the actual value of x with only 5 yes-no questions. Every such divide-and-conquer technique can be described as follows: First divide the set $A = \{1, 2, ..., 32\}$ into two parts A_{11} and A_{12} such that $|A_{11}| = |A_{12}|$ and

$$A_{11} \cap A_{12} = \Phi, \quad A_{11} \cup A_{12} = A.$$

Then ask Alice whether the value of x is in A_{11}. If the answer from Alice is "yes", then apply the same technique to A_{11}; otherwise apply the same technique to A_{12}. Repeating this technique fives times must give the unique answer. Here it is easily seen that the actual technique for dividing a set into two parts with the same number of elements is not important for this problem and there are many ways to do so. However, the crucial point of a divide-and-conquer technique for this problem is the difference between the numbers of the elements of the two subsets.

The optimality of the above divide-and-conquer techniques, by which we mean that each questioning gets the maximum amount of average information about the value of x, can be proved as follows. It is apparent that with one yes-no questioning the maximum amount of average information we can get is one bit. Since 32 is equal to a power of 2, i.e., $32 = 2^5$, it is possible to divide each new set containing the value of x into two parts with the same number of elements, and therefore with each questioning we get one bit of information about the value of x. On the other hand, the self information of the value of x is 5 bits. Thus, the above class of divide-and-conquer techniques is optimal.

We now analyze the general case $A = \{a_1, a_2, ..., a_n\}$, where $a_i \neq a_j$ for each pair (i, j) with $i \neq j$, that is, the value of x is one of A. Thus, the self information of x is $I(x) = \log_2 n$ bits. It is seen that any yes-no questioning which gets information about the value of x results in a partition $\{A_{11}, A_{12}\}$, that is,

$$A_{11} \cap A_{12} = \Phi, \quad A_{11} \cup A_{12} = A,$$

where A_{11} and A_{12} are nonempty. Thus, with one yes-no question the amount of average information we get from Alice is

$$I_{aver}(x; \text{ yes-no question}) = \frac{a}{n} \log_2 \frac{n}{a} + \frac{n-a}{n} \log_2 \frac{n}{n-a} \quad \text{bits},$$

where $a = |A_{11}|$. By elementary calculus the above average mutual information is maximal if and only if $a = n/2$ when n is even, and $a = \lfloor n/2 \rfloor$ or $\lfloor n/2 \rfloor - 1$ otherwise. Thus, with each questioning a divide-and-conquer technique should divide the set into two parts with the numbers of elements as equal as possible, in order to get the maximum amount of average information with each questioning. (See also Appendix A.)

Another example which can show the importance of the divide technique is the coin-and-balances problem. Given 13 coins, we are told that there is at most 1 false coin among them and that the false coin may be lighter or heavier than the genuine ones if there is a false coin. The problem is to design a scheme with which we can know whether there is a false coin and whether it is lighter or heavier than the genuine ones with the use of a beam balance only three times.

The example above shows the important role of the divide step of a divide-and-conquer technique. The conquer step is usually related to the divide step and also to the inherent complexity of the original problem. In what follows we study the relation between the CRA and some divide-and-conquer techniques.

Let $m_1, ..., m_n$ be n pairwise relatively prime positive integers, and $m = \prod_{i=1}^{n} m_i$. Recall that the CRT can be expressed as

$$\mathbf{Z}/(m) \cong \mathbf{Z}/(m_1) \times \cdots \times \mathbf{Z}/(m_n),$$

where the isomorphism is given by

$$\varphi : x \bmod m \to (x \bmod m_1, x \bmod m_2, \cdots, x \bmod m_n).$$

This theorem can be used to develop various kinds of divide-and-conquer techniques in algorithmics. With the CRA we can divide problems over $\mathbf{Z}/(m)$ into subproblems over $\mathbf{Z}/(m_i)$ (the divide procedure), then solve the subproblems and synthesize the solutions of the subproblems to get solutions of the original problem (conquer procedure). The technique, which is actually a parallel decomposition technique, can be summarized by Figure 4.1.

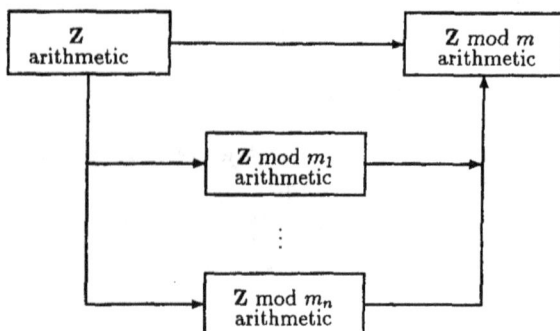

Figure 4.1: Divide-and-conquer techniques based on the CRA

As an example, we consider the multiplication in Z_m, which can be performed by doing multiplications modulo small integers. Note that the multiplication has quadratic computational complexity and

$$m^2 = m_1^2 \cdots m_n^2 > m_1^2 + \cdots + m_n^2,$$

indicating that such a transfer of a multiplication of $\mathbf{Z}/(m)$ is computationally worthwhile.

4.2   Polynomial Interpolation over Fields

The *polynomial interpolation problem* over a field can be described as follows: Given α_k, β_k for $k = 0, ..., n - 1$ over a field F (α_k's distinct), find $U(x) \in F[x]$ that satisfies

$$U(\alpha_k) = \beta_k, \quad k = 0, ..., n - 1.$$

For a polynomial $u(x) \in F[x]$ it is well known that

$$u(\alpha) = \beta \ \text{ if and only if } \ u(x) = \beta \bmod (x - \alpha).$$

Thus, the polynomial interpolation problem over F is nothing else but the following special Chinese Remainder Problem over $F[x]$: Find $U(x) \in F[x]$ such that

$$U(x) = \beta_k \bmod (x - \alpha_k), \quad k = 0, ..., n - 1. \tag{4.1}$$

To apply the general CRA to get an algorithm for the polynomial interpolation problem, we need the following lemma whose proof is trivial.

Lemma 4.2.1 *Let* $u(x) \in F[x], \alpha \in F$. *Then*

$$u(x) \bmod (x - \alpha) = u(\alpha)$$
$$u(x)^{-1} \bmod (x - \alpha) = u(\alpha)^{-1} \ if \ u(\alpha) \neq 0.$$

This lemma means that "polynomial remaindering" by $x - \alpha$ is equivalent to polynomial evaluation at $x = \alpha$.

Replacing mod $(x - \alpha)$ remaindering by evaluation at $x = \alpha$ and mod $(x - \alpha)$ inversion by evaluation at $x = \alpha$ followed by inversion in F in the iterative version of the general CRA, we obtain the following Newton's interpolation algorithm.

Newton's Interpolation Algorithm:

Step 0: (Initialization procedure) $U(x) = \beta_0$; $M(x) = 1$;

Step 1: (Repeat procedure) for $k := 1$ until $n - 1$ do
$M(x) := M(x) \times (x - \alpha_{k-1})$;
$c := M(\alpha_k)^{-1}$;
$\sigma := (\beta_k - U(\alpha_k)) \times c$;
$U(x) := U(x) + \sigma \times M(x)$

It is not difficult to prove that after the kth iteration Newton's interpolation algorithm computes

$$M(x) = M_k(x) = \prod_{i=0}^{k-1} (x - \alpha_i),$$
$$U(x) = U_k(x) = \sigma_0 M_0 + \cdots + \sigma_k M_k.$$

The solution $U(x)$ can be expressed in the form

$$\begin{aligned} U(x) &= \sigma_0 + \sigma_1(x - \alpha_0) + \sigma_2(x - \alpha_0)(x - \alpha_1) + \cdots \\ &+ \sigma_{n-1}(x - \alpha_0)(x - \alpha_1) \cdots (x - \alpha_{n-2}). \end{aligned}$$

Let F be a field with the field operation computable in time $O(1)$. Then for any n-point interpolation problem over F, Newton's algorithm computes the unique solution in $O(n^2)$ time. It should be noted that one of the most straightforward applications of the CRT is indeed the polynomial interpolation.

Another solution to the interpolation problem was given by Lagrange as

$$U(x) = \sum_{k=0}^{n-1} \beta_k L_k(x), \text{ where } L_k(x) = \prod_{i=0, \ i \neq k}^{n-1} \frac{x - \alpha_i}{\alpha_k - \alpha_i}.$$

The Lagrange algorithm is actually a special case of the CRA. To see this, define

$$n_i(x) = x - \alpha_i, \quad N(x) = \prod_{i=0}^{n-1}(x - \alpha_i), \quad N_i(x) = \frac{N(x)}{n_i(x)}.$$

Since α_i are pairwise distinct, the polynomials $(x - \alpha_i)$ are pairwise relatively prime. By definition the polynomials $N_i(x)$ and $n_i(x)$ are relatively prime, so there are two polynomials $s_i(x)$ and $t_i(x)$ over F such that

$$N_i(x)s_i(x) + n_i(x)t_i(x) = 1.$$

Hence

$$N_i(x)s_i(x) = 1 \bmod (x - \alpha_i),$$

from which it follows that

$$s_i(x) = N_i(\alpha_i)^{-1} \bmod (x - \alpha_i).$$

This proves Lemma 4.2.1. Finally, the CRA for the special Chinese Remainder Problem described in (4.1) becomes

$$
\begin{aligned}
U(x) &= \left[\sum_{i=0}^{n-1} \beta_i s_i(x) N_i(x) \bmod N(x)\right] \\
&= \sum_{i=0}^{n-1} \beta_i s_i(x) N_i(x) \\
&= \sum_{i=0}^{n-1} \beta_i \frac{N_i(x)}{N_i(\alpha_i)} \\
&= \sum_{i=0}^{n-1} \beta_i L_i(x),
\end{aligned}
$$

where $L_i(x)$ are defined as before.

Thus, both the Newton and Lagrange interpolation algorithms are special cases of the CRA. Specifically, the Newton interpolation algorithm corresponds to the iterative CRA, while the Lagrange interpolation algorithm corresponds to the CRA.

Let F be a field and $\alpha_1, \cdots, \alpha_n$ are n distinct nonzero elements of F. The *Vandermonde matrices* are defined as

$$
V(\alpha_i) =
\begin{bmatrix}
1 & \alpha_1^1 & \alpha_1^2 & \cdots & \alpha_1^{n-1} \\
1 & \alpha_2^1 & \alpha_2^2 & \cdots & \alpha_2^{n-1} \\
\vdots & \vdots & \vdots & \cdots & \vdots \\
1 & \alpha_n^1 & \alpha_n^2 & \cdots & \alpha_n^{n-1}
\end{bmatrix}. \tag{4.2}
$$

Vandermonde matrices have wide applications. An important problem concerning the Vandermonde matrices is the computation of their inverse matrices.

Since $\alpha_1, \cdots, \alpha_n$ are nonzero and distinct, the matrix of (4.2) is nonsingular. Let $A = [a_{ij}]$ denote the inverse of the Vandermonde matrix. By definition we get $V(\alpha_i)A = I_n$, where I_n is the $n \times n$ identity matrix. For each j define

$$a_j(x) = a_{1j} + a_{2j}x + a_{3j}x^2 + \cdots + a_{nj}x^{n-1}.$$

Thus, to compute the inverse matrix we need to find n polynomials $a_j(x)$ such that

$$a_j(\alpha_h) = \begin{cases} 0, & \text{if } h \neq j; \\ 1, & \text{otherwise.} \end{cases}$$

This can be done by the CRA directly or by the above two interpolation algorithms. The reader is invited to finish the rest of the computation of the Vandermonde matrix of (4.2).

4.3   Polynomial Interpolation over $\mathbf{Z}/(m)$

The polynomial interpolation problem over a field has been treated in the last section. Every function from a finite field F to F can be represented by a polynomial over F. However, this is not true for functions from $\mathbf{Z}/(m)$ to $\mathbf{Z}/(m)$. A simple example is the following. Consider the ring $\mathbf{Z}/(6)$ and the following function

$$
\begin{array}{c|cccccc}
x & 0 & 1 & 2 & 3 & 4 & 5 \\
f(x) & 0 & 2 & 3 & 1 & 4 & 5.
\end{array}
$$

It is easily seen there is no polynomial of $\mathbf{Z}/(m)[x]$ which represents the above function since $1 \bmod 2 = 3 \bmod 2 = 1$ but $f(1) \bmod 2 \neq f(3) \bmod 2$.

The *polynomial interpolation problem* over $\mathbf{Z}/(m)$ is the following. Given n distinct $\alpha_i \in \mathbf{Z}/(m)$ and n elements $\beta_i \in \mathbf{Z}/(m)$, find a polynomial $U(x) \in \mathbf{Z}/(m)[x]$ such that $U(\alpha_i) = \beta_i$ for all i if there is such a polynomial. In this section we give conditions for the existence of such a polynomial and an algorithm for solving the polynomial interpolation problem over $\mathbf{Z}/(m)$, where m is the product of distinct primes. The algorithm we describe is a combination of the CRA for integers and that for polynomials, and is an efficient one. One application of the algorithm is the construction of permutation polynomials over $\mathbf{Z}/(m)$. Another one is a secret-share scheme. Throughout this section we assume that $m = p_1 p_2 \cdots p_t$, where p_i are distinct primes and $p_1 < p_2 < \cdots < p_t$.

Necessary and sufficient conditions for the existence and uniqueness of a solution to the polynomial interpolation over $\mathbf{Z}/(m)$ are described by the following theorem.

Theorem 4.3.1 *Let α_i be n distinct elements of $\mathbf{Z}/(m)$, and $\beta_i \in \mathbf{Z}/(m)$, where $0 \leq i \leq n-1$. There is a polynomial $U(x) \in \mathbf{Z}/(m)[x]$ such that $U(\alpha_i) = \beta_i$ for all i with $0 \leq i \leq n-1$ if and only if α_i mod $p_j = \alpha_k$ mod p_j implies that β_i mod $p_j = \beta_k$ mod p_j for all possible i, j, k.*

Proof: Let φ be the mapping from $\mathbf{Z}/(m)$ to $\mathbf{Z}/(p_1) \times \cdots \times \mathbf{Z}/(p_t)$ given by

$$\varphi : x \bmod m \mapsto (x \bmod p_1, ..., x \bmod p_t).$$

By the CRT φ is an isomorphism. If there is a polynomial $U(x)$ described in the theorem, let $U(x) = u_0 + u_1 x + \cdots + u_{l-1} x^{l-1} \in \mathbf{Z}/(m)[x]$ and

$$U_i(x) = u_{i,0} + u_{i,1}x + \cdots + u_{i,l-1}x^{l-1},$$

where $u_{i,j} = u_j \bmod p_i$ for all i and j. It follows from $U(\alpha_i) = \beta_i$ for all i that $U_j(\alpha_i \bmod p_j) = \beta_i \bmod p_j$ for all i and j. Thus, if $\alpha_i \bmod p_j = \alpha_k \bmod p_j$, it must hold that $\beta_i \bmod p_j = \beta_k \bmod p_j$.

Conversely, if $\alpha_i \bmod p_j = \alpha_k \bmod p_j$ implies that $\beta_i \bmod p_j = \beta_k \bmod p_j$ for all possible i, j, k, we now construct a polynomial $U(x)$ as desired. For each j with $1 \leq j \leq t$, let $\alpha_{j,0}, \alpha_{j,1}, ..., \alpha_{j,n_j-1}$ be a set of all distinct elements of the set $\{\alpha_i \bmod p_j : i = 0, 1, ..., n-1\}$, and $\beta_{j,0}, \beta_{j,1}, ..., \beta_{j,n_j-1}$ be the corresponding elements of the set $\{\beta_i \bmod p_j : i = 0, 1, ..., n-1\}$, where $n_j \leq \min\{n, p_j\}$. As seen in the last section, the CRT for polynomials over a fields solves the polynomial interpolation problem over the fields $\mathbf{Z}/(p_i)$. Let

$$U_j(x) = u_{j,0} + u_{j,1}x + \cdots + u_{j,n_j-1}x^{n_j-1} \in \mathbf{Z}/(p_i)[x]$$

such that $U_j(\alpha_{j,k}) = \beta_{j,k}$ for $k = 0, 1, ..., n_j - 1$. Let $n' = \min\{n, p_t\}$, we define $u_{j,k} = 0$ for $k = n_j, n_j + 1, ..., n' - 1$. Thus, each $U_j(x)$ can be written as $U_j(x) = \sum_{k=0}^{n'-1} u_{j,k}x^k$. By the CRT and CRA we can compute integers u_k such that $u_k \bmod p_j = u_{j,k}$ for all j, where $0 \leq u_k \leq m-1$. Then $U(x) = \sum_{i=0}^{n'-1} u_i x^i$ is the polynomial of $\mathbf{Z}/(m)[x]$ such that $U(\alpha_i) = \beta_i$ for all i, where the degree of $U(x)$ is at most $n' - 1 \leq \min\{n-1, q_t - 1\}$. $\square$

Given t, n, $p_1, ..., p_t$, $\alpha_0, ..., \alpha_{n-1}$ (pairwise distinct), and $\beta_0, ..., \beta_{n-1} \in \mathbf{Z}/(m)$, a polynomial $U(x) \in \mathbf{Z}/(m)[x]$ such that $U(\alpha_i) = \beta_i$ for $i = 1, 2, ..., t$, if there exists one, is computed by the following algorithm.

The algorithm:

St0: Input t, n, $p_1, p_2, ..., p_t$, $\alpha_1, ..., \alpha_{n-1}$, and $\beta_1, ..., \beta_{n-1}$.

St1: Compute $\alpha_{j,i} = \alpha_i \bmod p_j$ and $\beta_{j,i} = \beta_i \bmod p_j$ for $i = 1, ..., t$ and $j = 0, 1, ..., n-1$.

St2: For $j = 1$ to t do the following. For each pair (i_1, i_2) with $0 \le i_1 < i_2 \le n - 1$ check whether $\alpha_{j,i_1} - \alpha_{j,i_2}$ is zero. If it is zero, check whether $\beta_{j,i_1} - \beta_{j,i_2}$ is zero. If $\beta_{j,i_1} - \beta_{j,i_2} \ne 0$, output "no solution".

St3: For $j = 1$ to t set n_j to be the number of distinct elements of the set $\{\alpha_{j,i} : i = 0, 1, ..., n - 1\}$, and set $\{a_{j,i} : i = 0, 1, ..., n_j - 1\}$ to be the n_j distinct elements and $\{b_{j,i} : i = 0, 1, ..., n_j - 1\}$ be the corresponding n_j distinct elements of $\{\beta_{j,i} : i = 0, 1, ..., n - 1\}$.

St4: For $j = 1$ to t do the following over $\mathbf{Z}/(p_j)$. Compute the polynomials

$$L_k(x) = \prod_{i=0,\ i \ne k}^{n_j - 1} \frac{x - a_i}{a_k - a_i}$$

for $k = 0, 1, ..., n_j - 1$, and then the polynomial $U_j(x)$ given by

$$U_j(x) = \sum_{k=0}^{n_j - 1} \beta_k L_k(x).$$

St5: For $j = 1$ to t set $u_{j,i}$ to be the coefficients of $U_j(x)$, i.e.,

$$U_j(x) = u_{j,0} + u_{j,1}x + \cdots + u_{j,n_j-1}x^{n_j-1},$$

and set $u_{j,k} = 0$ for $k = n_j, n_j + 1, ..., n' - 1$, where $n' = \max\{n, p_t\}$.

St6: Compute for $i = 1, 2, ..., t$

$$m = p_1 p_2 \cdots p_t$$

and $M_i = m/p_i$. For $i = 1$ to t use the Euclidean algorithm to compute two integers r_i and s_i such that $r_i p_i + s_i M_i = 1$. For $i = 0$ to $n' - 1$ compute

$$u_i = \sum_{j=1}^{t} s_i M_i u_{j,i} \bmod m.$$

St7: Output $U(x) = u_0 + u_1 x + \cdots + u_{n'-1}x^{n'-1}$.

The correctness of the algorithm has been shown by the proof of Theorem 4.3.1 and the CRA for integers as well as that for polynomials. Step 1 is the computation of the residues $\alpha_{j,i}$ and $\beta_{j,i}$, which needs tn modularization operations. Step 2 checks whether the polynomial interpolation problem for the given parameters has a solution, and it needs at most $2tn(n-1)$ comparisons for integers. Step 3 deletes the repeated elements in the sets $\{\alpha_{j,i} : i = 0, 1, ..., n-1\}$ and $\{\beta_{j,i} : i = 0, 1, ..., n-1\}$, and it needs at most $2tn(n-1)$ comparisons for integers. Step 4 solves polynomial interpolation problem over $\mathbf{Z}/(p_i)$ for $i = 1, ..., t$, and needs t calls of the special CRA for solving the polynomial interpolation problem over fields. Step 5 defines the values for some $u_{j,i}$, which is a preparation for the next step. Step 6 is the computation of the coefficients of the polynomial desired. In this step the CRA for integers is employed n times. Thus, the algorithm is actually a mixture of the CRA for integers and that for polynomials. Since the CRAs are efficient, so is the above algorithm. The above algorithm is based on the non-iterative version of the CRA. Another algorithm based on the iterative CRA can be similarly given.

Now we take an example. Let $m = 6 = 2 \times 3$, i.e., $p_1 = 2, p_2 = 3$. Given $(\alpha_0, \alpha_1, \alpha_2) = (0, 1, 2)$ and $(\beta_0, \beta_1, \beta_2) = (4, 5, 4)$. Step 1 computes $(\alpha_{1,0}, \alpha_{1,1}, \alpha_{1,2}) = (0, 1, 0)$ and $(\alpha_{2,0}, \alpha_{2,1}, \alpha_{2,2}) = (0, 1, 2)$, $(\beta_{1,0}, \beta_{1,1}, \beta_{1,2}) = (0, 1, 0)$ and $(\beta_{2,0}, \beta_{2,1}, \beta_{2,2}) = (1, 2, 1)$. Step 3 produces $(a_{1,0}, a_{1,1}) = (0, 1)$, $(b_{1,0}, b_{1,1}) = (0, 1)$, $(a_{2,0}, a_{2,1}, a_{2,2}) = (0, 1, 2)$, and $(b_{2,0}, b_{2,1}, b_{2,2}) = (1, 2, 1)$. With step 4 it is easy to get $U_1(x) = x$ and $U_2(x) = 2x^2 + 2x + 1$. With Steps 5 and 6 as well as 7 we get $U(x) = 2x^2 + 5x + 4$.

Polynomials over rings can be regarded both as functions or formal polynomials. Two polynomial functions $f(x)$ and $g(x)$ are said to be equal if $f(a) = g(a)$ for all a; while two formal polynomials are said to be equal if their corresponding coefficients are equal. This is an important distinction.

4.4   Shift-Register Synthesis over $\mathbf{Z}/(m)$

Let R be a commutative ring containing the unit element 1 and let R^* denote the set of invertible elements of R. A sequence $s^n = s_0 s_1 \cdots s_{n-1}$ over R is said to be generated by a *linear recurrence* of length l if there are elements $a_0 = 1, a_1, \cdots, a_l \in R$ such that

$$\sum_{i=0}^{l} a_i s_{j-i} = 0 \quad \text{for} \quad j = l, l+1, ..., n-1. \tag{4.3}$$

Such a linear recurrence of length l can be implemented with a *linear feedback shift-register* (briefly, *LFSR*) of length l over R, as depicted in Figure 4.2, where the boxes containing $s_{j-1}, s_{j-2}, \cdots, s_{j-l+1}, s_{j-l}$ are memory units, and at each time instant the content of the right-most memory unit is the output, while the contents of other memory units are shifted to their right-hand neighboring memory units respectively, and the left-most memory unit is then occupied by

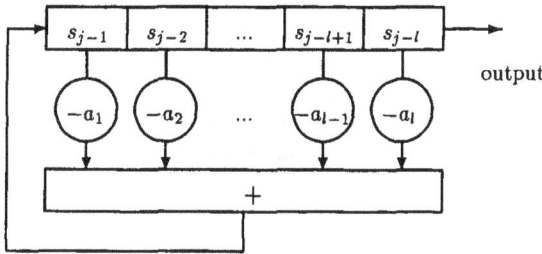

Figure 4.2: Linear feedback shift-registers

the new element

$$s_j = -a_1 s_{j-1} - \cdots - a_l s_{j-l}.$$

The polynomial $a(x) = a_0 + a_1 x + \cdots + a_l x^l$ is called the *connection polynomial* of the LFSR, and the sequence s^n is said to be produced by the LFSR. Linear-feedback shift-registers are a kind of the simplest finite automata and have wide applications in communications and cryptography [37, 30, 10].

Let $a(x) = a_0 + a_1 x + \cdots + a_l x^l$ and $s(x) = s_0 + s_1 x + \cdots + s_{n-1} x^{n-1}$. Then it is straightforward to see that

$$s(x)a(x) = \sum_{k=0}^{l+n-1} \left(\sum_{i+j=k,\ 0 \le i \le n-1,\ 0 \le j \le l} s_i a_j \right) x^k.$$

Thus, the linear recurrence relation (4.3) is equivalent to

$$\begin{cases} s(x)a(x) = b(x) \bmod x^n, \\ a(0) = 1, \end{cases} \tag{4.4}$$

for some polynomial $b(x) \in R[x]$ of degree $\le l - 1$. The length of the recurrence or LFSR satisfies

$$l \ge \max\{\deg a(x), 1 + \deg b(x)\}.$$

Without loss of generality we assume

$$l = \max\{\deg a(x), 1 + \deg b(x)\}.$$

For simplicity we write $A = (a(x), b(x))$ and denote

$$L(A) = \max\{\deg a(x), 1 + \deg b(x)\}.$$

By convention $\deg(0) = -\infty$.

The LFSR synthesis problem for sequences over a ring R is to find the shortest linear recurrence relation (4.3) for a given sequence over R or the shortest LFSR over R that generates a given sequence over R. There are a number of algorithms which solve the LFSR synthesis problem for sequences over fields, among them are the Berlekamp-Massey algorithm [8, 57], algorithms based on continued fractions and the Euclidean algorithm [26, 62, 91, 102]. Games and Chan [35] developed a fast algorithm for finding the shortest linear recurrence in the special case of a binary sequence of period 2^k. The Games-Chan algorithm was further generalized for sequences of period p^n over $GF(p^m)$ in [30].

Those algorithms for the LFSR synthesis of sequences over fields can be used to decode BCH and other codes [8, 57, 60, 51, 55, 66, 78, 91]. Since they are efficient it becomes necessary to take the length of the shortest LFSR that generates a sequence as a security measure for additive stream ciphers with this sequence as its keystream. In cryptography it is called the *linear span* or *linear complexity* of a sequence.

The algorithms mentioned above work only for sequences over fields. They don't work for sequences over $\mathbf{Z}/(m)$ when m is composite, since not all numbers have an inverse modulo m. An efficient algorithm for the LFSR synthesis of sequences over $\mathbf{Z}/(m)$ was developed by Reeds and Sloane [70]. The Reeds-Sloane algorithm, which has essentially the same complexity as the Berlekamp-Massey algorithm, can be used to decode BCH codes defined over integers modulo m [11, 12, 82, 87, 88] and makes it necessary to control the linear complexity of keystream sequences over $\mathbf{Z}/(m)$ for additive stream ciphering [86, 25]. The Reeds-Sloane algorithm is a typical divide-and-conquer algorithm based on the CRT.

In what follows the ring R is $\mathbf{Z}/(m)$, where $m \geq 2$ is a given integer whose factorization is known. The Reeds-Sloane algorithm is to find a linear recurrence $A = (a(x), b(x))$ that generates a given sequence over R, i.e., satisfies (4.4), and has minimal length $l = L(A)$. Let $m = \prod_{i=1}^{t} p_i^{e_i}, e_i \geq 1$, where the p_i's are distinct primes. The idea of this algorithm is first to divide the LFSR synthesis problem of a sequence s^n over $\mathbf{Z}/(m)$ into that of t sequences of the same length over the t rings $\mathbf{Z}/(p_i^{e_i})$'s for $i = 1, 2, ..., t$, i.e.,

$$s^n \bmod p_i^{e_i} = (s_j \bmod p_i^{e_i})_{j=0}^{n-1}, \quad i = 1, ..., t.$$

Then the conquer procedure includes the development of algorithm to synthesize the t sequences over $\mathbf{Z}/(p_i^{e_i})$'s respectively. Let $(a^{(i)}(x), b^i(x))$ denote the minimal linear recurrence of the sequence $s^n \bmod p_i^{e_i}$ for $i = 1, 2, ..., t$. Then the last step of the conquer procedure is to use the CRA to find a pair $(a(x), b(x))$ with

$$a(x) = a^{(i)}(x) \bmod p_i^{e_i}, \quad b(x) = b^{(i)}(x) \bmod p_i^{e_i}$$

for all $1 \leq i \leq t$, of length $l = \max\{l_i : 1 \leq i \leq t\}$. It is easy to show that $(a(x), b(x))$ is a minimal length recurrence generating the sequence s^n.

Let $RS(w^n, \mathbf{Z}/(p_i^{e_i}))$ denote the Reeds-Sloane algorithm for the LFSR synthesis of sequences w^n over $\mathbf{Z}/(p_i^{e_i})$, which will be described later. Then the general Reeds-Sloane algorithm can be described as follows.

The Reeds-Sloane algorithm:

Step 0: Input s^n.

Step 1: (Divide procedure) For $i = 1$ to t compute

$$s^n \bmod p_i^{e_i} = (s_j \bmod p_i^{e_i})_{j=0}^{n-1}.$$

Step 2: (Conquer procedure) For $i = 1$ to t compute

$$(a^{(i)}(x), b^i(x)) := RS(s^n \bmod p_i^{e_i}, \mathbf{Z}/(p_i^{e_i})).$$

Step 3: (Conquer procedure) Apply the polynomial CRA to $\{(a^{(i)}(x) : 1 \leq i \leq t\}$ to yield $a(x)$, and then apply the polynomial CRA to $\{(b^{(i)}(x) : 1 \leq i \leq t\}$ to yield $b(x)$.

Step 4: Output $(a(x), b(x))$.

With the help of the CRA, the LFSR synthesis problem for sequences over $\mathbf{Z}/(m)$ is converted into subproblems for sequences over $\mathbf{Z}/(p_i^{e_i})$. Given the sequence s^n over $\mathbf{Z}/(p^e)$, the special Reeds-Sloane algorithm is to find a linear recurrence $A = (a(x), b(x))$ of minimal length $l = L(A)$ satisfying (4.4). The key idea of the Reeds-Sloane algorithm is to consider not just (4.4) but the following more general problem.

Problem 4.4.1 *For all $\mu = 0, 1, ..., e - 1$, find pairs $A_\mu = (a_\mu(x), b_\mu(x))$ such that*

$$\begin{cases} s(x)a_\mu(x) = b_\mu(x) \bmod x^n, \\ a_\mu(0) = p^\mu, \end{cases} \tag{4.5}$$

and $L(A_\mu) = l_\mu$ is minimized.

The Reeds-Sloane algorithm is an iterative procedure that, for all $0 \leq k \leq n, 0 \leq \mu < e$, calculates pairs $A_\mu^{(k)} = (a_\mu^{(k)}(x), b_\mu^{(k)}(x))$ satisfying

$$\begin{cases} s(x)a_\mu^{(k)}(x) = b_\mu^{(k)}(x) \bmod x^k, \\ a_\mu^{(k)}(0) = p^\mu, \end{cases}$$

and minimizing $L(A_\mu^{(k)})$. It is an extension of the Berlekamp-Massey algorithm in the following sense. Let $p^{u_{\mu k}}$, where $0 \le u_{\mu k} \le e$, be the highest power of p dividing the coefficient of x^k in

$$s(x)a_\mu^{(k)}(x) - b_\mu^{(k)}(x)$$

if this coefficient is nonzero, otherwise $u_{\mu k}$ is set to be equal to e. Then at the kth step in the iteration of the Reeds-Sloane algorithm the following property holds for all $0 \le r < k$:

(P_r) For all $0 \le g < e$, either

$$L(A_g^{(r+1)}) = L(A_g^{(r)}) \tag{4.6}$$

or else there exists an $h = f(g, r)$ with

$$\begin{aligned} &g + u_{hr} < e, \\ &L(A_g^{(r+1)}) = r + 1 - L(A_h^{(r)}), \\ &L(A_g^{(r+1)}) > L(A_g^{(r)}). \end{aligned} \tag{4.7}$$

This property is the analogue of the conditions that Berlekamp gives in [8, p.183, eq. (7.314)] and Massey gives in [57, p.123, eqs. (11)-(13).]. Given this data, the Reeds-Sloane algorithm calculates $A_\mu^{(k+1)}$ and $f(\mu, k)$, where $0 \le \mu < e$, such that property (P_k) holds. The quantity $L(A_\mu^{(k)})$ satisfies also the inequalities

$$L(A_{\mu+1}^{(k)}) \le L(A_\mu^{(k)}) \le L(A_\mu^{(k+1)}). \tag{4.8}$$

Given s^n over $\mathbf{Z}/(p^e)$, the Reeds-Sloane algorithm produces a pair $A = (a(x), b(x))$ such that $s(x)a(x) = b(x) \bmod x^n$, where $a(0) = 1$, and the length $l = L(A) = \max\{\deg a(x), 1 + \deg b(x)\}$ is minimized. The algorithm can be described as follows.

The algorithm $RS(s^n, \mathbf{Z}/(p^e))$:

Sp 2.0: For each $\mu = 0, 1, ..., e - 1$ set

$$a_\mu(x) = p^\mu, b_\mu(x) = 0, a_\mu^{new}(x) = p^\mu, b_\mu^{new}(x) = p^\mu s_0,$$

and find θ_μ and u_μ, where $\theta_\mu \in (\mathbf{Z}/(p^e))^*$ and $0 \le u_\mu \le e$, such that

$$s(x)a_\mu(x) - b_\mu(x) = \theta_\mu p^{u_\mu} \bmod x.$$

Sp $2.k$: (This step is carried out for each $k = 1, 2, ..., n - 1$) There are three parts within this step:

First, for each $g = 0, 1, ..., e - 1$, if

$$L(a_g^{new}(x), b_g^{new}(x)) > L(a_g(x), b_g(x)),$$

set

$$
\begin{aligned}
a_g^{old}(x) &= a_h(x), & u_g^{old} &= u_h, \\
b_g^{old}(x) &= b_h(x), & r_g &= k - 1, \\
\theta_g^{old} &= \theta_h,
\end{aligned}
$$

where $h = e - 1 - u_g$.

Second, for each $\mu = 0, 1, ..., e - 1$, set $a_\mu(x) = a_\mu^{new}(x)$ and $b_\mu(x) = b_\mu^{new}(x)$.

Third, for each $\mu = 0, 1, ..., e - 1$, find θ_μ and u_μ, where $\theta_\mu \in (\mathbf{Z}/(p^e))^*$ and $0 \leq u_\mu \leq e$, such that

$$s(x)a_\mu(x) - b_\mu(x) = \theta_\mu p^{u_\mu} x^k \bmod x^{k+1}.$$

Set $g = e - 1 - u_\mu$. Then

1. if $u_\mu = e$ set

$$a_\mu^{new}(x) = a_\mu(x), \quad b_\mu^{new}(x) = b_\mu(x);$$

2. if $u_\mu \neq e$ and $L(a_g(x), b_g(x)) = 0$, set

$$a_\mu^{new}(x) = a_\mu(x), \quad b_\mu^{new}(x) = b_\mu(x) + \theta_\mu p^{u_\mu} x^k;$$

3. if $u_\mu \neq e$ and $L(a_g(x), b_g(x)) \neq 0$, set

$$
\begin{aligned}
a_\mu^{new}(x) &= a_\mu(x) - \theta_\mu(\theta_g^{old})^{-1} p^{u_\mu - u_g^{old}} x^{k-r_g} a_g^{old}(x), \\
b_\mu^{new}(x) &= b_\mu(x) - \theta_\mu(\theta_g^{old})^{-1} p^{u_\mu - u_g^{old}} x^{k-r_g} b_g^{old}(x).
\end{aligned}
$$

At the end of Step $n - 1$ the Reeds-Sloane algorithm terminates and a minimal linear recurrence $(a(x), b(x))$ is given by $(a_0^{new}(x), a_0^{new}(x))$. If the modulus m is fixed, $O(n^2)$ steps are required to synthesize a sequence of length n [70]. Changing from a sequence modulo p to a sequence modulo p^e increases the number of steps by a factor e. For the proof of the correctness of the Reeds-Sloane algorithm, we refer to Reeds and Sloane [70]. It should be noted that the CRA makes the divide-and-conquer technique possible.

4.5   Common Primitive Roots

For every $n \geq 1$, Euler's function $\Phi(n)$ is defined to be the number of integers a, $1 \leq a < n$ such that $\gcd(a, n) = 1$. This function has the following properties:

1. If $n = p$ is a prime, then $\Phi(p) = p - 1$.

2. For any prime p, $\Phi(p^k) = p^{k-1}(p - 1)$.

3. If $m, n \geq 1$ and $\gcd(m, n) = 1$, then $\Phi(mn) = \Phi(m)\Phi(n)$.

4. For any integer $n = \prod_p p^k$, $\Phi(n) = \prod_p p^{k-1}(p - 1)$.

Let q be an integer with $\gcd(q, n) = 1$. Then there must be an integer d such that $q^d = 1 \bmod n$, since the group $(\mathbf{Z}/(n))^*$ is finite. The smallest such a positive integer d is called the *order* of q modulo n, denoted by $\mathrm{ord}_n(q)$. An integer q is said to be a *primitive root* of (or modulo) n if $\mathrm{ord}_n(q) = \Phi(n)$. For $g = g' \bmod N$, g is a primitive root of N if and only if g' is a primitive root of N. The investigation of primitive roots dates back to Gauss who proved the following result [36].

Proposition 4.5.1 *If p is a prime, then there exist $\Phi(p-1)$ primitive roots of p. The only integers having primitive roots are p^e, $2p^e$, 1, 2 and 4, with p an odd prime and $e \geq 1$.*

Generally, if m has a primitive root g then m has exactly $\Phi(\Phi(m))$ primitive roots and they are given by the numbers in the set

$$S = \{g^j : 1 \leq j \leq \Phi(m), \text{ and } \gcd(j, \Phi(m)) = 1\}.$$

The proof of this conclusion is straightforward. Here we count the primitive roots of m only in the range from 2 to $m - 1$.

The distribution of primitive roots has been investigated by, e.g., Carlitz [20], Vegh [94, 95, 96], Szalay [92] and Shoup [84]. Investigations in the least primitive roots have been carried out by Burgess and Elliott [18], Elliott [32], Wang [98], Heath-Brown [40] and Murata [64].

It is shown in [31] that primitive roots are quite useful in constructing cryptographic keystream sequences. A integer g is called a *common primitive root* of a set of integers $n_1, n_2, ..., n_t$ if it is a primitive root of each n_i for $i = 1, ..., t$. Common primitive roots are related to generalized cyclotomy and twin difference sets [101]. The most interesting application of common primitive roots may be in stream ciphering [31].

The CRT shows not only the existence of common primitive roots, but also how to find a common primitive root, as described by the following theorem.

Theorem 4.5.2 *Let g_i be a primitive root of n_i for $i = 1, ..., t$, where n_i are pairwise relatively prime. Then the unique solution g of*

$$g = g_i \bmod n_i \quad for \ i = 1, 2, ..., t.$$

is a common primitive root of n_i.

Proof: The existence and uniqueness of the above g is guaranteed by the CRT. It is straightforward to see that g is a primitive root of each n_i. □

Thus, we can use the CRA to compute a common primitive root of $\{n_i : 1 \le i \le t\}$ if we know a primitive root of each n_i. With the CRT we can prove the following result, which is also useful in cryptography [31].

Theorem 4.5.3 *Let $n_1, n_2, ..., n_t$ be pairwise relatively prime positive integers, and g an integer with $\gcd(g, n_i) = 1$ for each $1 \le i \le t$. Then*

$$\mathrm{ord}_{n_1 n_2 \cdots n_t}(g) = \mathrm{lcm}\{\mathrm{ord}_{n_1}(g), \mathrm{ord}_{n_2}(g), ..., \mathrm{ord}_{n_t}(g)\}.$$

Proof: By the CRT we have

$$\mathbf{Z}/(n) \cong \mathbf{Z}/(n_1) \times \cdots \times \mathbf{Z}/(n_t),$$

where $n = n_1 \cdots n_t$, and the isomorphism is given by

$$\varphi(x \bmod n) = (x \bmod n_1, ..., x \bmod n_t).$$

If $g^d = 1 \bmod n$, then $\varphi(g^d \bmod n) = \varphi(1)$. It follows that

$$(g^d \bmod n_1, ..., g^d \bmod n_t) = (1, ..., 1).$$

Thus, each $\mathrm{ord}_{n_i}(g)$ divides d, and therefore

$$\mathrm{lcm}\{\mathrm{ord}_{n_1}(g), ..., \mathrm{ord}_{n_t}(g)\} | \mathrm{ord}_{n_1 \cdots n_t}(g).$$

On the other hand, let $d = \mathrm{lcm}\{\mathrm{ord}_{n_1}(g), ..., \mathrm{ord}_{n_t}(g)\}$. Then it is easy to see that $\varphi(g^d \bmod n) = \varphi(1) = (1, ..., 1)$. Thus, $g^d \bmod n = 1$. It follows that $\mathrm{ord}_{n_1 n_2 \cdots n_t}(g)$ must be equal to d. □

Theorem 4.5.3 can be used to establish a divide-and-conquer method for the computation of the order of a modulo a composite integer, which can be briefly described as follows. First develop an algorithm for the computation of the order of a modulo p^k. Let

$$n = p_1^{e_1} p_2^{e_2} \cdots p_t^{e_t},$$

where p_i are distinct primes. Then compute $\mathrm{ord}_{p_i^{e_i}}(a)$ for each i. Finally, compute the least common multiple of these $\mathrm{ord}_{p_i^{e_i}}(a)$. This gives the required result.

4.6   From One- to Multi-dimension

In digital signal processing one-dimensional data are often rearranged into multi-dimensional arrays for the purpose of achieving parallel computation. Such a re-arrangement of data often makes it possible to develop fast algorithms [68, 1, 2].

Let $A = \{a[0], a[1], ..., a[n-1]\}$ be a one-dimensional array of data. Such data is often said to be an n-*point array*. Assume $n = n_1 n_2 \cdots n_k$ and let

$$\mu : \mathbf{Z}/(n) \longrightarrow \mathbf{Z}/(n_1) \times \cdots \times \mathbf{Z}/(n_k)$$

be a *bijection*. Such a bijection induces a data rearrangement with which a one-dimensional array $A = \{a[i]\}$ having n points can be rearranged into an k-dimensional array $\underline{A} = \{\underline{a}[i_1, ..., i_k]\}$ defined by

$$\underline{a}[i_1, ..., i_k] = a[\mu^{-1}(i_1, ..., i_k)] \tag{4.9}$$

for each $(i_1, ..., i_k) \in \mathbf{Z}/(n_1) \times \cdots \times \mathbf{Z}/(n_k)$. Since there are $n!$ such bijections, there are $n! - 1$ possible rearrangements. Some of the $n!$ bijections are homomorphisms of the two rings, others are not. Some of them are *one-way* in the sense that $\mu(x)$ is easy to compute when μ and x are given, but it is hard to find an $x \in \mathbf{Z}/(n)$ such that $\mu(x) = y$ when μ and $y \in \mathbf{Z}/(n_1) \times \cdots \times \mathbf{Z}/(n_k)$ are given; others are *two-way* in the sense that the computation of $\mu(x)$ and the finding of an x satisfying $\mu(x) = y$ both are easy. Such a one-way bijection could be useful in cryptography, but may be of little value in fast computation. In contrast two-way homomorphisms are very popular in fast computations, but might be awful in some cryptographic applications.

Let $n = \prod_{i=1}^{k} n_i$, where n_i are pairwise relatively prime. The mapping

$$\varphi(x) = (x \bmod n_1, ..., x \bmod n_k)$$

is a ring homomorphism between the two rings $\mathbf{Z}/(n)$ and $\mathbf{Z}/(n_1) \times \cdots \times \mathbf{Z}/(n_k)$. In addition, the CRA shows that the mapping is a two-way bijection. In digital computation the data rearrangement of (4.9) based on the mapping φ plays an important role. We shall refer to this as the *CRT data rearrangement*.

For the CRT data rearrangement the original data indices can be recovered by the CRA. For $1 \le i \le k$ let $N_i = n/n_i$. With the Euclidean algorithm we can compute integers u_i such that

$$N_i u_i = 1 \bmod n_i.$$

Thus the CRA computes the original one-dimensional indices from the k-dimensional indices $(r_1, ..., r_k)$ by

$$\varphi^{-1}(r_1, ..., r_k) = \sum_{i=1}^{k} r_i u_i N_i \bmod n.$$

The CRT data rearrangement is illustrated by an example. Consider the case $n = 6$. Taking $n_1 = 2$ and $n_3 = 3$, we have the inverse mapping

$$\varphi^{-1}(r_1, r_2) = 3r_1 + 4r_2 \bmod 6.$$

Therefore, a one-dimensional 6-point array $A = \{a[0], a[1], ..., a[5]\}$ is rearranged into a two-dimensional array

$$\begin{bmatrix} \underline{a}[0,0] & \underline{a}[0,1] & \underline{a}[0,2] \\ \underline{a}[1,0] & \underline{a}[1,1] & \underline{a}[1,2] \end{bmatrix} = \begin{bmatrix} a[0] & a[4] & a[2] \\ a[3] & a[1] & a[5] \end{bmatrix}.$$

If n is a power of a prime, the CRT data rearrangement technique does not work. In this case there are other possible data rearrangements. For example, let $n = p^k$. Then each integer i with $0 \le i \le n - 1$ can be expressed as

$$i = i_0 + i_1 p + \cdots + i_{k-1} p^{k-1},$$

where $0 \le i_j \le p - 1$ for each $0 \le j \le k - 1$. By defining

$$\underline{a}[i_0, ..., i_{k-1}] = a[i_0 + i_1 p + \cdots + i_{k-1} p^{k-1}]$$

a one-dimensional array $A = \{a[0], a[1], ..., a[n - 1]\}$ is rearranged into a k-dimensional array $\underline{A} = \{\underline{a}[i_0, ..., i_{k-1}]\}$. However, the bijection

$$\mu(i_0, ..., i_{k-1}) = i_0 + i_1 p + \cdots + i_{k-1} p^{k-1}$$

is not a homomorphism between the two rings $\mathbf{Z}/(p^k)$ and $\mathbf{Z}/(p) \times \cdots \times \mathbf{Z}/(p)$.

Another useful data rearrangement technique is the so-called *lexicographic rearrangement* [1]. Let $n = n_1 n_2$ and $A = \{a[i]\}$ be an n-point data array. The two-dimensional lexicographic rearrangement changes A into a two-dimensional array $\underline{A}$ by

$$\underline{a}[i, j] = a[n_2 i + j],$$

where i ranges from 0 to $n_1 - 1$ and j from 0 to $n_2 - 1$. It is readily seen that the mapping

$$\mu(n_2 i + j) = (i, j)$$

is injective. It is not hard to generalize the two-dimensional lexicographic rearrangement into an k-dimensional one.

Consider now the case $n = 6$, $n_1 = 2$, and $n_2 = 3$. The two-dimensional lexicographic rearrangement gives

$$\begin{bmatrix} \underline{a}[0,0] & \underline{a}[0,1] & \underline{a}[0,2] \\ \underline{a}[1,0] & \underline{a}[1,1] & \underline{a}[1,2] \end{bmatrix} = \begin{bmatrix} a[0] & a[1] & a[2] \\ a[3] & a[4] & a[5] \end{bmatrix}.$$

Another data rearrangement technique is based on the generalized CRT described by Theorem 2.5.1. Let $n = \prod_{i=1}^{k} n_i$, where n_i are pairwise relatively prime. Let a_i be integers such that

$$\gcd(a_i, n_i) = 1, \quad i = 1, 2, ..., k.$$

With the mapping

$$\theta(x) = (a_1 x \bmod n_1, ..., a_k x \bmod n_k)$$

one-dimensional data $A = \{a[i]\}$ is converted into a k-dimensional array

$$\underline{A} = \{\underline{a}[r_1, ..., r_k]\},$$

where $r_j = r a_j \bmod n_j$ for $j = 1, 2, ..., k$. Let $N_i = n/n_i$ for $i = 1, 2, ..., k$. With the Euclidean algorithm we can compute c_i such that $c_i a_i = 1 \bmod n_i$. Then the generalized CRA computes the original one-dimensional indices from the k-dimensional indices $(r_1, r_2, ..., r_k)$ as

$$\theta^{-1}(r_1, ..., r_k) = \prod_{i=1}^{k} N_i u_i c_i r_i \bmod n_i.$$

The following example shows the above two-way homomorphic data rearrangement. Let $n_1 = 3, n_2 = 4, a_1 = 2, a_2 = 3$. Then $n = n_1 n_2 = 12, c_1 = 2, c_2 = 3$. A one-dimensional 12-point data $A = \{a[i]\}$ is rearranged into the following two-dimensional array:

$$
\begin{bmatrix}
\underline{a}[0,0] & \underline{a}[0,1] & \underline{a}[0,2] & \underline{a}[0,3] \\
\underline{a}[1,0] & \underline{a}[1,1] & \underline{a}[1,2] & \underline{a}[1,3] \\
\underline{a}[2,0] & \underline{a}[2,1] & \underline{a}[2,2] & \underline{a}[2,3]
\end{bmatrix}
=
\begin{bmatrix}
a[0] & a[3] & a[6] & a[9] \\
a[8] & a[11] & a[2] & a[5] \\
a[4] & a[7] & a[10] & a[1]
\end{bmatrix}.
$$

This rearrangement technique based on the generalized CRA is a generalization of the one based on the CRA, and it will be used in a fast Fourier transform later.

It is worth noting that the residue class rings $\mathbf{Z}/(p^k)$ cannot be further expressed as a product of two residue class rings $\mathbf{Z}/(m_1)$ and $\mathbf{Z}/(m_2)$ homomorphically, i.e., there are no rings $\mathbf{Z}/(m_1)$ and $\mathbf{Z}/(m_2)$ such that their direct product is isomorphic to $\mathbf{Z}/(p^k)$, where p is a prime. This fact may limit the application of the data rearrangement techniques based on the CRA and generalized CRAs.

4.7 A Modular Algorithm for Cyclic Convolution

The *cyclic convolution* of two n-point arrays

$$\{a_0, a_1, ..., a_{n-1}\} \text{ and } \{b_0, b_1, ..., b_{n-1}\}$$

is defined by

$$c_i = \sum_{k=0}^{n-1} a_{(i-k) \bmod n} b_k, \quad i = 0, 1, ..., n-1, \tag{4.10}$$

where a_i and b_i belong to a field F. In matrix form we have

$$
\begin{bmatrix}
c_0 \\
c_1 \\
c_2 \\
\vdots \\
c_{n-1}
\end{bmatrix}
=
\begin{bmatrix}
a_0 & a_{n-1} & \cdots & a_2 & a_1 \\
a_1 & a_0 & \cdots & a_3 & a_2 \\
a_2 & a_1 & \cdots & a_4 & a_3 \\
\vdots & \vdots & \vdots & \vdots & \vdots \\
a_{n-1} & a_{n-2} & \cdots & a_1 & a_0
\end{bmatrix}
\begin{bmatrix}
b_0 \\
b_1 \\
b_2 \\
\vdots \\
b_{n-1}
\end{bmatrix}. \tag{4.11}
$$

It is easily seen that (4.11) is the system of coefficients of the polynomial

$$c(x) = a(x)b(x) \bmod (x^n - 1), \tag{4.12}$$

where

$$
\begin{aligned}
a(x) &= a_0 + a_1 x + \cdots + a_{n-1} x^{n-1}, \\
b(x) &= b_0 + b_1 x + \cdots + b_{n-1} x^{n-1}, \\
c(x) &= c_0 + c_1 x + \cdots + c_{n-1} x^{n-1}.
\end{aligned}
$$

Thus the computation of the cyclic convolution of (4.11) is converted into that of the polynomial multiplication modulo $(x^n - 1)$.

Let $m_0(x), m_1(x), ..., m_{k-1}(x)$ be pairwise relatively prime polynomials over the field F such that

$$x^n - 1 = m_0(x)m_1(x) \cdots m_{k-1}(x).$$

To compute the $a(x)b(x) \bmod (x^n - 1)$, we apply the general modular technique outlined in Section 3.1. We first compute the polynomials

$$c_i(x) = a(x)b(x) \bmod m_i(x), \quad i = 0, 1, ..., k-1.$$

Then by the CRA for polynomials the polynomial $c(x) = a(x)b(x) \bmod (x^n - 1)$ can be recovered by using

$$c(x) = \sum_{i=0}^{k-1} c_i(x)u_i(x)M_i(x) \bmod (x^n - 1),$$

where

$$M_i(x) = \frac{x^n - 1}{m_i(x)}$$

and $u_i(x)$ is the polynomial satisfying

$$u_i(x)M_i(x) = 1 \bmod m_i(x).$$

We refer to this algorithm as the *Winograd modular algorithm for cyclic convolution* [103].

At this moment we have a good opportunity to demonstrate the power of modular techniques based on the CRA. Consider now the above algorithm for the case $n = 3$ and $F = GF(2)$. Note that

$$x^3 - 1 = (x - 1)(x^2 + x + 1).$$

We take the moduli $m_0(x) = x - 1$ and $m_1(x) = x^2 + x + 1$ over $GF(2)$. Then it is easily checked that

$$
\begin{aligned}
c_0(x) &= [a(x)b(x) \bmod (x - 1)] \\
&= [(a_0 + a_1 x + a_2 x^2)(b_0 + b_1 x + b_2 x^2) \bmod (x - 1)] \\
&= (a_0 + a_1 + a_2)(b_0 + b_1 + b_2).
\end{aligned}
$$

Set

$$
\begin{aligned}
h_0 &= (a_0 + a_1 + a_2)(b_0 + b_1 + b_2), \\
h_1 &= (a_0 + a_2)(b_0 + b_2), \\
h_2 &= (a_1 + a_2)(b_1 + b_2), \\
h_3 &= (a_0 + a_1)(b_0 + b_1).
\end{aligned}
$$

Then the polynomial $c_1(x)$ can be expressed as

$$
\begin{aligned}
c_1(x) &= [a(x)b(x) \bmod (x^2 + x + 1)] \\
&= (h_1 + h_2) + (h_1 + h_3)x.
\end{aligned}
$$

Since $m_1(x) - xm_0(x) = 1$, with the CRA for polynomials we can recover $c(x)$ by

$$
\begin{aligned}
c(x) &= [c_0(x)m_1(x) - xc_1(x)m_0(x) \bmod (x^3 - 1)] \\
&= [h_0(x^2 + x + 1) + x(x + 1)[(h_1 + h_2) + (h_1 + h_3)x] \bmod (x^3 - 1)] \\
&= (h_0 + h_2 + h_3)x^2 + (h_0 + h_1 + h_2)x + (h_0 + h_1 + h_3).
\end{aligned}
$$

Setting $s = (h_0 + h_2)$, we obtain

$$c(x) = (s + h_3)x^2 + (s + h_1)x + (h_0 + h_1 + h_3).$$

Thus, with this modular algorithm the computation of the 3-point cyclic convolution needs only 4 multiplications and 13 additions. This means that only $\lfloor n \log_2 n \rfloor$ multiplications are needed. In contrast, the direct calculation based on (4.11) needs $n^2 = 9$ multiplications and 6 additions.

Notice that this method is not applicable for some cyclic convolutions. For example, since $x^{2^n} - 1 = (x - 1)^{2^n}$ over $GF(2^m)$, the 2^n-point cyclic convolution over $GF(2^m)$ cannot be performed in this way.

4.8 A Fast Algorithm for Cyclic Convolution

In Section 4.7 we have seen that the Winograd modular algorithm could be quite fast if the length n of the data arrays is small. This algorithm can be combined with the CRT data rearrangement technique to obtain another fast algorithm for cyclic convolution [103].

Let $\{a_i\}$ and $\{b_i\}$ be two n-point data arrays over a field F. Recall that the cyclic convolution is defined by (4.10), we repeat it here for convenience:

$$c_i = \sum_{k=0}^{n-1} a_{(i-k) \bmod n} b_k, \quad i = 0, 1, ..., n - 1. \tag{4.13}$$

Let $n = n_1 n_2$, where n_1 and n_2 are relatively prime. The CRT data rearrangement replaces the one-dimensional indices i and k with the two-dimensional indices (i_1, i_2) and (k_1, k_2), where

$$(i_1, i_2) = (i \bmod n_1, i \bmod n_2),$$
$$(k_1, k_2) = (k \bmod n_1, k \bmod n_2).$$

Since n_1 and n_2 are relatively prime, with the Euclidean algorithm we get two integers u_1 and u_2 such that

$$u_1 n_1 + u_2 n_2 = 1.$$

Then the original one-dimensional indices i and k are recovered from the two-dimensional indices by

$$i = u_2 n_2 i_1 + u_1 n_1 i_2 \bmod n,$$
$$k = u_2 n_2 k_1 + u_1 n_1 k_2 \bmod n.$$

With this CRT data rearrangement the convolution (4.13) is converted into

$$c_{i_1,i_2} = \sum_{k_1=0}^{n_1-1} \sum_{k_2=0}^{n_2-1} a_{(i_1-k_1,i_2-k_2)\mathrm{mod}\,n} b_{k_1,k_2}, \tag{4.14}$$

where the two-dimensional arrays $\{c_{i_1,i_2}\}$, $\{a_{i_1,i_2}\}$, and $\{b_{i_1,i_2}\}$, with indices $i_1 = 0, ..., n_1 - 1$ and $i_2 = 0, ..., n_2 - 1$, are defined by

$$b_{i_1,i_2} = b_{(u_2 n_2 i_1 + u_1 n_1 i_2)\mathrm{mod}\,n},$$
$$a_{i_1,i_2} = a_{(u_2 n_2 i_1 + u_1 n_1 i_2)\mathrm{mod}\,n},$$
$$c_{i_1,i_2} = c_{(u_2 n_2 i_1 + u_1 n_1 i_2)\mathrm{mod}\,n}.$$

Converting (4.14) back to the original indices, we have

$$c_{(u_2 n_2 i_1 + u_1 n_1 i_2)} = \sum_{k_1=0}^{n_1-1} \sum_{k_2=0}^{n_2-1} a_{(u_2 n_2 (i_1-k_1)+u_1 n_1 (i_2-k_2))} b_{(u_2 n_2 k_1 + u_1 n_1 k_2)},$$

where the indices are understood modulo n. Thus the n-point cyclic convolution has been converted into a two-dimensional cyclic convolution. This will result in an efficient algorithm only if efficient n_1- and n_2-point cyclic convolution algorithms are available. Using the Winograd modular algorithm for cyclic convolution described in Section 4.7 for the n_1- and n_2-point cyclic convolution together with the CRT data rearrangement gives another fast algorithm for cyclic convolution due to Winograd [103].

To illustrate this algorithm, we consider the case $n = 6 = 2 \times 3$ and $F = GF(2)$. In this case the cyclic convolution

$$
\begin{aligned}
c(x) &= c_0 + c_5 x + c_4 x^2 c_3 x^3 + c_2 x^4 + c_1 x^5 \bmod (x^6 - 1) \\
&= (a_0 + a_5 x + a_4 x^2 + a_3 x^2 + c_2 x^4 + a_1 x^5)(\sum_{i=0}^{5} b_i x^i) \bmod (x^6 - 1)
\end{aligned}
$$

can be written as

$$
\begin{bmatrix} c_0 \\ c_1 \\ c_2 \\ c_3 \\ c_4 \\ c_5 \end{bmatrix} =
\begin{bmatrix}
a_0 & a_1 & a_2 & a_3 & a_4 & a_5 \\
a_1 & a_2 & a_3 & a_4 & a_5 & a_0 \\
a_2 & a_3 & a_4 & a_5 & a_0 & a_1 \\
a_3 & a_4 & a_5 & a_0 & a_1 & a_2 \\
a_4 & a_5 & a_0 & a_1 & a_2 & a_3 \\
a_5 & a_0 & a_1 & a_2 & a_3 & a_4
\end{bmatrix}
\begin{bmatrix} b_0 \\ b_1 \\ b_2 \\ b_3 \\ b_4 \\ b_5 \end{bmatrix}. \tag{4.15}
$$

Taking $n_1 = 2$ and $n_2 = 3$ and applying the CRT data rearrangement, we get the following two dimensional convolution

$$
\begin{bmatrix} c_{0,0} \\ c_{0,1} \\ c_{0,2} \\ c_{1,0} \\ c_{1,1} \\ c_{1,2} \end{bmatrix}
=
\begin{bmatrix}
a_{0,0} & a_{0,1} & a_{0,2} & a_{1,0} & a_{1,1} & a_{1.2} \\
a_{0,1} & a_{0,2} & a_{0,0} & a_{1,1} & a_{1,2} & a_{1,0} \\
a_{0,2} & a_{0,0} & a_{0,1} & a_{1,2} & a_{1,0} & a_{1,1} \\
a_{1,0} & a_{1,1} & a_{1.2} & a_{0,0} & a_{0,1} & a_{0,2} \\
a_{1,1} & a_{1,2} & a_{1,0} & a_{0,1} & a_{0,2} & a_{0,0} \\
a_{1,2} & a_{1,0} & a_{1,1} & a_{0,2} & a_{0,0} & a_{0,1}
\end{bmatrix}
\begin{bmatrix} b_{0,0} \\ b_{0,1} \\ b_{0,2} \\ b_{1,0} \\ b_{1,1} \\ b_{1,2} \end{bmatrix}.
\tag{4.16}
$$

Converting (4.16) back to the original one-dimensional indices by the CRA, we obtain that

$$
\begin{bmatrix} c_0 \\ c_4 \\ c_2 \\ c_3 \\ c_1 \\ c_5 \end{bmatrix}
=
\begin{bmatrix}
a_0 & a_4 & a_2 & a_3 & a_1 & a_5 \\
a_4 & a_2 & a_0 & a_1 & a_5 & a_3 \\
a_2 & a_0 & a_4 & a_5 & a_3 & a_1 \\
a_3 & a_1 & a_5 & a_0 & a_4 & a_2 \\
a_1 & a_5 & a_3 & a_4 & a_2 & a_0 \\
a_5 & a_3 & a_1 & a_2 & a_0 & a_4
\end{bmatrix}
\begin{bmatrix} b_0 \\ b_4 \\ b_2 \\ b_3 \\ b_1 \\ b_5 \end{bmatrix}.
\tag{4.17}
$$

Set

$$
C_0 = \begin{bmatrix} c_0 \\ c_4 \\ c_2 \end{bmatrix}, \quad
C_1 = \begin{bmatrix} c_3 \\ c_1 \\ c_5 \end{bmatrix}, \quad
B_0 = \begin{bmatrix} b_0 \\ b_4 \\ b_2 \end{bmatrix}, \quad
B_1 = \begin{bmatrix} b_3 \\ b_1 \\ b_5 \end{bmatrix}
$$

and

$$
A_0 = \begin{bmatrix} a_0 & a_4 & a_2 \\ a_4 & a_2 & a_0 \\ a_2 & a_0 & a_4 \end{bmatrix}, \quad
A_1 = \begin{bmatrix} a_3 & a_1 & a_5 \\ a_1 & a_5 & a_3 \\ a_5 & a_3 & a_1 \end{bmatrix}.
$$

Then (4.17) can be written as

$$
\begin{bmatrix} C_0 \\ C_1 \end{bmatrix}
=
\begin{bmatrix} A_0 & A_1 \\ A_1 & A_0 \end{bmatrix}
\begin{bmatrix} B_0 \\ B_1 \end{bmatrix},
$$

which is equivalent to

$$
\begin{aligned}
C_0 &= A_0 B_0 + A_1 B_1, \\
C_1 &= A_1 B_0 + A_0 B_1.
\end{aligned}
$$

Here $A_0 B_0$, $A_1 B_1$, $A_1 B_0$ and $A_0 B_1$ are 3-point cyclic convolutions. If we apply directly the 3-point modular algorithm described in Section 4.7, we need to invoke it four times. There is a trick for reducing the number of times for invoking the 3-point cyclic convolution algorithm. To this end, define the 3-point cyclic convolution

$$C = (A_0 + A_1)(B_0 + B_1).$$

Then $C_1 = C - C_0$. Thus, we need to invoke the 3-point algorithm twice for the calculation of C_0, and one time for that of C. Since the 3-point algorithm needs 4 multiplications and 13 additions, this algorithm based on the 3-point algorithm and the CRT data rearrangement needs 12 multiplications and 57 additions, while the brute-force calculation based on (4.15) requires 36 multiplications and 24 additions.

4.9   Fast Fourier Transform and CRT

Let F be a field and $\alpha \in F$ be an nth primitive root of unity, i.e., $\alpha^n = 1$, but $\alpha^k \neq 1$ for each k with $1 \leq k \leq n - 1$. Let $(a_0, a_1, ..., a_{n-1}) \in F^n$. The *discrete Fourier transform* (briefly, DFT) is defined by

$$A_k = \sum_{i=0}^{n-1} \alpha^{ik} a_i, \quad k = 0, 1, ..., n - 1. \tag{4.18}$$

Let

$$\mathbf{A} = \begin{bmatrix} A_0 \\ A_1 \\ \vdots \\ A_{n-1} \end{bmatrix}, \quad \mathbf{a} = \begin{bmatrix} a_0 \\ a_1 \\ \vdots \\ a_{n-1} \end{bmatrix}.$$

In the matrix form the discrete Fourier transform is

$$\mathbf{A} = W\mathbf{a}, \tag{4.19}$$

where W is the $n \times n$ matrix with entries $w_{i,j} = \alpha^{ij}$. Clearly, the DFT transforms a vector $\mathbf{a}$ into another vector $\mathbf{A}$. Assume that $n \cdot 1 \neq 0$, where 1 denotes the unity of the field F, then the inverse Fourier transform is given by

$$a_k = \frac{1}{n} \sum_{i=0}^{n-1} \alpha^{-ik} A_i, \quad k = 0, 1, ..., n - 1. \tag{4.20}$$

In the matrix form the inverse Fourier transform is

$$\mathbf{a} = W^{-1}\mathbf{A}, \tag{4.21}$$

where W^{-1} denotes the inverse of the matrix W. Thus, the inverse Fourier transform has the same form as the Fourier transform. A fast Fourier transform (FFT) is a fast algorithm for the calculation of a discrete Fourier transform.

There are now quite a number of fast algorithms for the calculation of the discrete Fourier transform. Among them are the Cooley-Tukey FFT [23, 9], the Good-Thomas FFT [38, 93], and the Winograd FFT [103]. The Cooley-Tukey FFT, one of the best FFTs in certain cases, is based on the CRT. Its basic idea is to break a one-dimensional Fourier transform of block length $n_1 n_2$ into a two-dimensional Fourier transform, where n_1 and n_2 are relatively prime. The data rearrangement techniques in the Good-Thomas FFT are based on the CRT.

The Good-Thomas FFT can be derived as follows. Let $n = n_1 n_2$, where n_1 and n_2 are relatively prime. We again compute two integers u_1 and u_2 such that

$$u_1 n_1 + u_2 n_2 = 1.$$

Recall that the CRT data rearrangement replaces the one-dimensional index i with the two-dimensional index (i_1, i_2), where

$$(i_1, i_2) = (i \bmod n_1, i \bmod n_2). \tag{4.22}$$

The original one-dimensional indices are recovered from the two-dimensional ones by

$$i = u_2 n_2 i_1 + u_1 n_1 i_2.$$

There is a variant of the CRT data rearrangement, i.e., the one based on the generalized CRT described by Theorem 2.5.1. This variant replaces the one-dimensional index k by the two-dimensional one (k_1, k_2), where

$$(k_1, k_2) = (u_2 k \bmod n_1, u_1 k \bmod n_2), \quad \gcd(u_2, n_1) = \gcd(u_1, n_2) = 1. \tag{4.23}$$

It is clear that the two-dimensional indices defined by this variant are a permutation of those defined by the CRT data rearrangement. To make it more clear, let $(i_1, i_2)_1$ denote the two-dimensional index defined by the CRT data rearrangement, and $(i_1, i_2)_2$ defined by its variant described above. Then we have clearly

$$(i_1, i_2)_2 = (u_2, u_1)(i_1, i_2)_1$$

and

$$(i_1, i_2)_1 = (n_2, n_1)(i_1, i_2)_2.$$

In this variant of the CRT data rearrangement the original indices k are recovered by

$$k = n_2 k_1 + n_1 k_2.$$

In certain applications this variant is more convenient than the original CRT data rearrangement.

In the Good-Thomas FFT the input indices i are rearranged by the CRT data rearrangement formulated in (4.22), while the output indices are rearranged by its variant described by (4.23). With these data rearrangements the one-dimensional DFT becomes

$$
\begin{aligned}
A_{k_1,k_2} &= \sum_{i_1=0}^{n_1-1} \sum_{i_2=0}^{n_2-1} \alpha^{u_2 n_2^2 i_1 k_1} \alpha^{u_1 n_1^2 i_2 k_2} a_{i_1,i_2} \\
&= \sum_{i_1=0}^{n_1-1} \beta^{i_1 k_1} \sum_{i_2=0}^{n_2-1} \gamma^{i_2 k_2} a_{i_1,i_2},
\end{aligned}
\tag{4.24}
$$

where $\beta = \alpha^{u_2 n_2^2}$ and $\gamma = \alpha^{u_1 n_1^2}$. It is easily verified that β and γ are n_1th and n_2th primitive root of unity respectively.

Converting back to the original indices we have

$$
A_{n_2 k_1 + n_1 k_2} = \sum_{i_1=0}^{n_1-1} \sum_{i_2=0}^{n_2-1} \alpha^{(u_2 n_2 i_1 + u_1 n_1 i_2)(n_2 k_1 + n_1 k_2)} a_{u_2 n_2 i_1 + u_1 n_1 i_2},
$$

where the indices are reduced by modulo n.

With the above data rearrangements the one-dimensional DFT is converted into a two-dimensional one described by (4.24). If n_1 or n_2 is the product of two relatively prime integers, then the two-dimensional DFT can be further split into a multidimensional DFT. Then we can use other FFTs to compute the small DFTs or DFTs with block length of a power of a prime.

To illustrate the Good-Thomas FFT, we consider the case $n = 6$. Taking $n_1 = 2$, $n_3 = 3$, and α as a 6th primitive root of unity, we have

$$u_1 = -1, \quad u_2 = 1, \quad \beta = -1, \quad \gamma = \alpha^2,$$

where u_1, u_2, β and γ are defined as before. Thus, the two-dimensional DFT of (4.24) becomes

$$
\begin{aligned}
A_{k_1,k_2} &= \sum_{i_1=0}^{1} \sum_{i_2=0}^{2} \gamma^{i_2 k_2} a_{i_1,i_2} \\
&= \sum_{i_2=0}^{2} \gamma^{i_2 k_2} (a_{0,i_2} + a_{1,i_2}).
\end{aligned}
$$

Thus, we know $A_{0,k_2} = A_{1,k_2}$ for all $k_2 = 0, 1, 2$. Furthermore, we have

$$
\begin{aligned}
A_{k_1,0} &= x + y + z, \\
A_{k_1,1} &= x - z\alpha + y\alpha^2, \\
A_{k_1,2} &= x - y\alpha + z\alpha^2,
\end{aligned}
\tag{4.25}
$$

where

$$
x = a_{0,0} + a_{1,0}, \quad y = a_{0,1} + a_{1,1}, \quad z = a_{0,2} + a_{1,2}.
\tag{4.26}
$$

It is clear from (4.25) and (4.26) that we need only five multiplications and nine additions of the field F to calculate the original 6-point DFT.

To make a comparison between the above algorithm and classical algorithms for the DFT, let us take the classical algorithm for the calculation of the DFT based on polynomial evaluation. Let

$$
a(x) = a_0 + a_1 x + \cdots a_{n-1} x^{n-1},
$$

where the vector $(a_0, ..., a_{n-1})$ is the input vector of the DFT. Then it is clear that the DFT can be expressed as

$$
A_k = a(\alpha^k), \quad k = 0, ..., n - 1.
$$

Thus, the calculation of the DFT becomes the evaluation of the polynomial $a(x)$ at the points $1, \alpha, \alpha^2, ..., \alpha^{n-1}$. There is a clever way for the polynomial evaluation which can be described by

$$
a(x) = ((\cdots ((a_{n-1} x + a_{n-2})x + a_{n-3}) \cdots)x + a_1)x + a_0.
$$

With this recursive procedure the evaluation of $a(x)$ at a point x needs $n - 1$ multiplications and $n - 1$ additions of the field F, and therefore the calculation of the DFT needs $(n - 1)^2$ multiplications and $(n - 1)n$ additions, since A_0 does not need a multiplication. With this approach the calculation of the 6-point DFT needs 26 multiplications and 30 additions; while the above algorithm needs only 5 multiplications and 9 additions. This clearly shows the difference.

We now point out a very important relation between CRT and the discrete Fourier transform. Let notations be the same as before, and let $a(x) = a_0 + a_1 x + \cdots + a_{n-1} x^{n-1}$. The discrete Fourier transform becomes

$$
A_k = a(\alpha^k) = a(x) \bmod m_i(x), \quad k = 0, 1, ..., n - 1,
$$

where $m_i(x) = x - \alpha^i$ for all i. Thus, the DFT is a special CRT. Let $m(x) = \prod_{i=0}^{n-1} m_i(x) = x^n - 1$, and $M_i(x) = m(x)/m_i(x)$. By the CRA it is easy to see that

$$
a(x) = \sum_{i=0}^{n-1} \frac{M_i(x)}{M_i(\alpha^i)} A_i.
\tag{4.27}
$$

Thus, the inverse DFT is in fact a special CRA. By refining (4.27) it is possible to get a FFT.

A slight generalization of the discrete Fourier transform is the following. Let $\alpha_0, \alpha_1, \cdots, \alpha_{n-1}$ be n distinct nonzero elements of a field F. The generalized DFT transforms a vector $(a_0, a_1, \cdots, a_{n-1}) \in F^n$ into another vector $(A_0, A_1, \cdots, A_{n-1})$, where

$$A_j = \sum_{i=0}^{n-1} a_i \alpha_j^i, \quad j = 0, 1, \cdots, n-1.$$

The inverse transform is given by the CRA or the two polynomial interpolation formulae.

The most general transform technique may be the CRT transform which includes quite a number of famous transform techniques as special cases. The CRA is the inverse of the CRT transform.

Chapter 5

In Bridging Computations

Matters in the rings $\mathbf{Z}/(m)$ look much more complicated than those in some fields, since some elements of the rings $\mathbf{Z}/(m)$ have zero divisors when m is composite. An important application of the CRT and CRA is the bridging of phenomena over $\mathbf{Z}/(m)$ with those over some finite fields. This will be the topic of this chapter. It should be noted that the applications considered in this chapter are only partial, and that further applications are possible. One advantage of the rings $\mathbf{Z}/(m)$ over many finite fields is that matters over the former have a numerical realization which makes a fast implementation possible.

5.1 A Main Bridge

One bridge between rings $\mathbf{Z}/(m)$ and fields $Z/(p)$ is supported by the CRT, and described by the following theorem.

Theorem 5.1.1 *Let* $f(x_1, ..., x_n) \in \mathbf{Z}/(m)[x_1, ..., x_n]$, *where* $m = m_1 \cdots m_t$ *and* m_i *are pairwise relatively prime. Define*

$$f_i(x_1, ..., x_n) = f(x_1, ..., x_n) \bmod m_i \in \mathbf{Z}/(m_i)[x_1, ..., x_n], \quad i = 1, ..., t$$

and let $N(f)$ *and* $N(f_i)$ *denote the number of zeros of* f *over* $\mathbf{Z}/(m)$ *and* f_i *over* $\mathbf{Z}/(m_i)$ *respectively. Then*

$$N(f) = N(f_1)N(f_2) \cdots N(f_t).$$

Proof: Let ϕ be the mapping from $\mathbf{Z}/(m)$ to $\mathbf{Z}/(m_1) \times \cdots \times \mathbf{Z}/(m_t)$ given by

$$\phi : x \mapsto (x \bmod m_1, ..., x \bmod m_t).$$

By the CRT ϕ is an isomorphism. Let

$$f(x_1, ..., x_n) = \sum_{(e_1, ..., e_n)} a_{e_1 e_2 \cdots e_n} x_1^{e_1} x_2^{e_2} \cdots x_n^{e_n},$$

95

where $a_{e_1 \cdots e_n} \in \mathbf{Z}/(m)$ and $e_i \geq 0$ are nonnegative integers. A natural extension of ϕ is given by

$$\phi(f) = \sum_{(e_1, \ldots, e_n)} \phi[a_{e_1 e_2 \cdots e_n}] x_1^{e_1} x_2^{e_2} \cdots x_n^{e_n}.$$

Thus, the extended ϕ is an isomorphism between the rings $\mathbf{Z}/(m)[x_1, \ldots, x_n]$ and $\mathbf{Z}/(m_1)[x_1, \ldots, x_n] \times \cdots \times \mathbf{Z}/(m_t)[x_1, \ldots, x_n]$.

By the assumption and definition f_i and f are polynomials in n variables. If $(x_1^{(i)}, \ldots, x_n^{(i)}) \in (\mathbf{Z}/(m_i))^n$ is a solution of $f_i = 0$ for $i = 1, \ldots, t$, by definition the vector

$$(\phi^{-1}(x_1^{(1)}, \ldots, x_1^{(t)}), \ldots, \phi^{-1}(x_n^{(1)}, \ldots, x_n^{(t)})) \in (\mathbf{Z}/(m))^n$$

is a solution of $f = 0$. By the CRT $N(f) \geq N(f_1)N(f_2) \cdots N(f_t)$.

On the other hand, if $(x_1, \ldots, x_n) \in (\mathbf{Z}/(m))^n$ is a solution of $f = 0$, let $x_j^{(i)} = x_j \bmod m_i$ for all i and j. Then by definition $(x_1^{(i)}, \ldots, x_n^{(i)}) \in (\mathbf{Z}/(m_i))^n$ is a solution of $f_i = 0$ for $i = 1, \ldots, t$. Thus, $N(f) \leq N(f_1)N(f_2) \cdots N(f_t)$. Combining the above two inequalities proves the theorem. $\square$

With the help of this theorem computing the number of zeros of a polynomial in n variables over $\mathbf{Z}/(m)$ becomes that of t polynomials over $\mathbf{Z}/(m_i)$. On the other hand, let $M_i = m/m_i$. Since M_i and m_i are relative prime, with Euclidean algorithm we compute two integers u_i and v_i such that $u_i M_i + m_i v_i = 1$. Let $(x_1^{(i)}, \ldots, x_n^{(i)})$ be a zero of f_i over $\mathbf{Z}/(m_i)$ for $i = 1, \ldots, t$. Then the CRA gives a zero of f as

$$\left(\sum_{i=1}^t M_i u_i x_1^{(i)} \bmod m, \ldots, \sum_{i=1}^t M_i u_i x_n^{(i)} \bmod m \right).$$

This computes all the zeros of f over $\mathbf{Z}/(m)$ when $(x_1^{(i)}, \ldots, x_n^{(i)})$ ranges over all the solutions of $f_i = 0$ over $\mathbf{Z}/(m_i)$. Thus, the CRT and CRA establish a bridge between the zeros of f over $\mathbf{Z}/(m)$ and those of f_i over $\mathbf{Z}/(m_i)$.

The most important case is when m_i are primes. Then many results concerning some finite fields can be transferred to those over the rings $\mathbf{Z}/(m_1 \cdots m_t)$. In addition, many problems relating to the number of solutions of equations over $\mathbf{Z}/(m)$ could also be solved with the help of Theorem 5.1.1.

We end this section with an example. Consider the equation $x^2 + y^2 = 1$ over $\mathbf{Z}/(6)$. It is easily examined that the equation

$$x^2 + y^2 = 1$$

has the solutions $(1,0)$, $(0,1)$ over $\mathbf{Z}/(2)$, and the solutions $(0,1)$, $(0,2)$, $(1,0)$, $(2,0)$ over $\mathbf{Z}/(3)$. By the CRA this equation has the following solutions over $\mathbf{Z}/(6)$: $(3,4)$, $(3,2)$, $(1,0)$, $(5,0)$, $(0,1)$, $(0,5)$, $(4,3)$, $(2,3)$.

5.2 Solving Equations over Z/(m)

In this section we consider only some special equations over $\mathbf{Z}/(m)$, where $m = p_1 p_2 \cdots p_t$. The purpose is to show the importance of the CRT and CRA in solving equations. Throughout this section, $p, p_1, p_2, ..., p_t$ are distinct primes, and $m = p_1 p_2 \cdots p_t$.

Our first equation is $x^2 + y^2 = 1$ over $\mathbf{Z}/(m)$. To count the number of solutions of this equation over $\mathbf{Z}/(m)$, we first consider it over $Z/(p)$. Let $p > 2$ be a prime and a an integer not divisible by p. Then the integer a is called a *quadratic residue* modulo p if there is an integer x such that $x^2 = a \bmod p$, and a *quadratic nonresidue* otherwise. The *Legendre symbol* $\left(\frac{a}{p}\right)$ is defined to be 1 if a is a quadratic residue modulo p, -1 if it is a quadratic nonresidue, and 0 otherwise (that is, p divides a). Let $N_m(f)$ denote the number of zeros of a polynomial f over $\mathbf{Z}_m$. Since $N_p(x^2 - a) = 1 + \left(\frac{a}{p}\right)$, we get that

$$
\begin{aligned}
N_p(x^2 + y^2 - 1) &= \sum_{a+b=1} N(x^2 - a)N(y^2 - b) \\
&= p + \sum_a \left(\frac{a}{p}\right) + \sum_b \left(\frac{b}{p}\right) + \sum_{a+b=1} \left(\frac{a}{p}\right)\left(\frac{b}{p}\right),
\end{aligned}
$$

where $a, b \in Z/(p)$. It is easily seen that

$$
\sum_a \left(\frac{a}{p}\right) = 0, \quad \left(\frac{a^2}{p}\right) = 1, \ a \neq 0 \bmod p.
$$

Setting $c = a^{-1}$ for $a \neq 0 \bmod p$, we obtain that

$$
\begin{aligned}
\sum_{a+b=1} \left(\frac{a}{p}\right)\left(\frac{b}{p}\right) &= \sum_{a \in Z/(p)} \left(\frac{a(1-a)}{p}\right) \\
&= \sum_{c \in Z/(p)^\bullet} \left(\frac{c-1}{p}\right)\left(\frac{(c^{-1})^2}{p}\right) \\
&= \sum_{c \in Z/(p)^\bullet} \left(\frac{c-1}{p}\right) \\
&= -\left(\frac{-1}{p}\right) + \sum_{c \in Z/(p)} \left(\frac{c-1}{p}\right) \\
&= -\left(\frac{-1}{p}\right) = -(-1)^{(p-1)/2} \bmod p \\
&= \begin{cases} -1, & \text{if } p = 1 \bmod 4, \\ +1, & \text{if } p = 3 \bmod 4, \end{cases}
\end{aligned}
$$

Combining the above results yields

$$N_p(x^2 + y^2 - 1) = p - \left(\frac{-1}{p}\right).$$

By this formula and Theorem 5.1.1 we obtain the following conclusion.

Theorem 5.2.1 *Let the symbols be as before. Then the number of solutions of the equation $x^2 + y^2 = 1$ over $\mathbf{Z}/(m)$ is given by*

$$N_m(x^2 + y^2 - 1) = \prod_{i=1}^{t}\left(p_i - \left(\frac{-1}{p_i}\right)\right).$$

In his *Disquisitiones arithmeticae* (§358), by introducing cyclotomic numbers Gauss proved that if $p = 1 \bmod 3$, then there are integers a and b such that $a = 1 \bmod 3$ and $4p = a^2 + 27b^2$, and

$$N_p(x^3 + y^3 - 1) = p - 2 + a.$$

From this result and Theorem 5.1.1 we get the following theorem.

Theorem 5.2.2 *If the primes p_i are of the form $3j + 1$, let $4p_i = a_i^2 + 27b_i^2$, where $a_i = 1 \bmod 3$. Then the number of solutions of the equation $x^3 + y^3 = 1$ over $\mathbf{Z}/(m)$ is given by*

$$N_m(x^3 + y^3 - 1) = \prod_{i=1}^{t}(p_i - 2 + a_i).$$

With the help of character sums it is proved (see [43, p.102]) that if r is odd, then the number of solutions of the equation $x_1^2 + x_2^2 + \cdots + x_r^2 = 1$ over $\mathbf{Z}/(p)$ is given by

$$N_p(x_1^2 + x_2^2 + \cdots + x_r^2 - 1) = p^{r-1} + (-1)^{((r-1)/2)((p-1)/2)}p^{(r-1)/2}.$$

If r is even, then

$$N_p(x_1^2 + x_2^2 + \cdots + x_r^2 - 1) = p^{r-1} + (-1)^{(r/2)((p-1)/2)}p^{r/2-1}.$$

Thus, the conclusion of the following theorem follows from Theorem 5.1.1.

Theorem 5.2.3 *If r is odd, then the number of solutions of the equation $x_1^2 + x_2^2 + \cdots + x_r^2 = 1$ over $\mathbf{Z}/(m)$ is given by*

$$N_m(x_1^2 + x_2^2 + \cdots + x_r^2 - 1) = \prod_{i=1}^{t}[p_i^{r-1} + (-1)^{((r-1)/2)((p_i-1)/2)}p_i^{(r-1)/2}].$$

If r is even, then

$$N_m(x_1^2 + x_2^2 + \cdots + x_r^2 - 1) = \prod_{i=1}^{t}[p_i^{r-1} + (-1)^{(r/2)((p_i-1)/2)}p_i^{r/2-1}].$$

If the exact solutions of $f(x_1, ..., x_n)$ over $\mathbf{Z}/(m)$ are required, they can be obtained from those of $f_i(x_1, ..., x_n)$ over $\mathbf{Z}/(p_i)$ with the help of the CRA. Let $P_i = m/p_i$, then $\gcd(P_i, p_i) = 1$. With Euclidean algorithm we compute two integers u_i and v_i such that $p_i v_i + P_i u_i = 1$. Let $(x_1^{(i)}, ..., x_n^{(i)})$ be a solution of $f_i(x_1, ..., x_n)$ over $\mathbf{Z}/(p_i)$ for $i = 1, ..., t$. Then

$$\left(\sum_{i=1}^{t} P_i u_i x_1^{(i)} \bmod m, ..., \sum_{i=1}^{t} P_i u_i x_n^{(i)} \bmod m \right)$$

is a solution of $f(x_1, ..., x_n)$ over $\mathbf{Z}/(m)$. All of the solutions of f are obtained when $(x_1^{(i)}, ..., x_n^{(i)})$ ranges over all the solutions of $\mathbf{Z}/(p_i)$.

Finally, we consider the following set of linear equations over $\mathbf{Z}/(m)$

$$a_{i,1}x_1 + a_{i,2}x_2 + \cdots + a_{i,n}x_n = b_i, \quad i = 1, 2, \cdots, k. \tag{5.1}$$

Set the $k \times n$ matrix $A = [a_{i,j}]$, $B = (b_1, ..., b_k)$, $A_h = [a_{i,j} \bmod p_h]$, $B_h = (b_1 \bmod p_h, ..., b_k \bmod p_h)$ for $h = 1, ..., t$. Then the equation (5.1) has solutions over $\mathbf{Z}/(m)$ if and only if each equation

$$A_i X^T = B_i^T \tag{5.2}$$

has solutions over $\mathbf{Z}/(p_i)$, where $X = (x_1, ..., x_n)$ and X^T denotes the transpose of the vector. Since the t linear equations are over fields, one can easily give necessary and sufficient conditions for the solvability of these sets of linear equations. By solving the t set of equations over fields, we can obtain all solutions of the equation (5.1) from those of (5.2) with the CRA, as described above. Let s_i be the number of solutions of the equation (5.2), it is then easily seen that the number of solutions of the equation (5.1) is $s_1 s_2 \cdots s_t$.

5.3   Number of Roots of Equations over $\mathbf{Z}/(m)$

A basic question about polynomials in n variables over $\mathbf{Z}/(m)$ is when they have a zero in $(\mathbf{Z}/(m))^n$. Here and hereafter in this section $m = p_1 p_2 \cdots p_t$ and p_i are pairwise distinct primes. We shall answer this question with the help of the main bridge of Theorem 5.1.1 supported by the CRT.

If $f \in \mathbf{Z}/(m)[x_1, ..., x_n]$, then it can be expressed as

$$f(x) = \sum_{(i_1, i_2, ..., i_n)} a_{i_1 i_2 \cdots i_n} x_1^{i_1} x_2^{i_2} \cdots x_n^{i_n},$$

where the sum is over a finite number of n-tuples of nonnegative integers $(i_1, i_2, ..., i_n)$, and where $a_{i_1 i_2 \cdots i_n} \neq 0$. A polynomial of the form $x_1^{i_1} x_2^{i_2} \cdots x_n^{i_n}$

is called a *monomial*. Its total degree is defined to be $i_1 + i_2 + \cdots i_n$, and its degree in the variable x_k is defined as i_k. The degree of $f(x)$ is the maximum of the total degrees of monomials with nonzero coefficients that occur in $f(x)$. The degree in x_k is defined similarly. It is easily seen that

$$\deg(fg) \leq \deg(f) + \deg(g),$$

where $\deg(f)$ denotes the degree of $f(x)$. The equality does not necessarily hold here, since some elements of $\mathbf{Z}/(m)$ have zero divisors.

Let F be a field and $f(x) \in F[x_1, ..., x_n]$. Suppose that

1. $f(0, 0, ..., 0) = 0$,

2. $n > \deg(f)$.

It can be shown that $f(x)$ has at least two zeros in F^n [43, p.143]. Combining this conclusion with Theorem 5.1.1 proves the following theorem.

Theorem 5.3.1 *Let $f(x) \in \mathbf{Z}/(m)[x_1, ..., x_n]$ and suppose that $f(0, 0, ..., 0) = 0$ and $n > \deg(f_i)$, where $f_i = f \bmod p_i$ as before. Then $f(x)$ has at least 2^t zeros in $(\mathbf{Z}/(m))^n$.*

To examine this result, we consider the following example. Let $m = 2 \times 3$, $f(x) = 2x_1 + 3x_2$. Then $f_1(x_1, x_2) = x_2$, $n = 2 > \deg(f_1) = 1$, and its zeros over $\mathbf{Z}/(2)$ are $(0, 0), (1, 0)$. Similarly, $f_2(x_1, x_2) = 2x_1$, $n = 2 > \deg(f_2) = 1$, and its zeros over $\mathbf{Z}/(3)$ are $(0, 0), (0, 1), (0, 3)$. Thus, by the CRA all the zeros of f are $(0, 0), (0, 4), (0, 2), (3, 0), (3, 4), (3, 2)$. Thus, the number of solutions of f over $\mathbf{Z}/(6)$ is greater than $2^2 = 4$.

It should be noted that the condition that $n > \deg(f_i)$ is not necessary. The example $f(x) = 2x_1^2 + 3x_2^3$ can show this.

An important result about zeros of polynomials over finite fields is the so-called Riemann Hypothesis for Curves over Finite Fields [100, 79, p.2]. It can be stated as follows. Suppose that $f(x, y)$ is a polynomial of degree d, with coefficients in $GF(q)$, and that $f(x, y)$ is absolutely irreducible, i.e., irreducible not only over $GF(q)$, but also over every algebraic extension thereof. Then

$$|N(f) - q| \leq 2g\sqrt{q} + c(d),$$

where g is the "genus" of the curve $f(x, y) = 0$ and where $c(d)$ is a constant depending on d. It can be proved that $g \leq (d-1)(d-2)/2$. Thus,

$$|N(f) - q| \leq (d-1)(d-2)\sqrt{q} + c(d). \tag{5.3}$$

Combining equation (5.3) with Theorem 5.1.1 yields the following conclusion.

Theorem 5.3.2 *Let $f(x,y) \in \mathbf{Z}/(m)[x,y]$, $f_i(x,y) = f(x,y) \bmod p_i$, and $d_i = \deg(f_i)$. Suppose that each polynomial $f_i(x,y)$ is absolutely irreducible. Then*

$$\prod_i [p_i - (d_i - 1)(d_i - 2)\sqrt{p_i} - c_i(d_i)] \leq N(f)$$

$$\prod_i [p_i + (d_i - 1)(d_i - 2)\sqrt{p_i} + c_i(d_i)] \geq N(f)$$

where $N(f)$ is the number of zeros of $f(x,y)$ over $\mathbf{Z}/(m)$, and where $c_i(d_i)$ is a constant depending on d_i.

For some polynomials it is easy to check whether they are absolutely irreducible. For instance, the polynomial $f(x,y) = y^d - F(x) \in Z/(p)[x,y]$ is absolutely irreducible if $\gcd(\deg(F), d) = 1$ [79, p.13]. The above result of Theorem 5.3.2 might be considered as an analog of the Riemann Hypothesis for Curves over Finite Fields.

We take the above two theorems only as examples about the theory of equations over $\mathbf{Z}/(m)$, based on the main bridge of Theorem 5.1.1 supported by the CRT. Most of the results about equations over finite fields can be transferred to $\mathbf{Z}/(m)$ directly, where $m = p_1 p_2 \cdots p_t$ as before. To illustrate this further, we take another example. Let $f(x_1, ..., x_n) \in GF(q)[x_1, ..., x_n]$ be a nonzero polynomial of degree d. It is proved in [79, p.147] that the number of zeros of $f(x_1, ..., x_n)$ satisfies

$$N(f) \leq dq^{n-1}.$$

Combining this with Theorem 5.1.1 gives the following.

Theorem 5.3.3 *Let $f(x_1, ..., x_n) \in \mathbf{Z}/(m)[x_1, ..., x_n]$, and for $i = 1, ..., t$ let*

$$f_i(x_1, ..., x_n) = f(x_1, ..., x_n) \bmod p_i,$$

and $d_i = \deg(f_i)$. Suppose that f_i are nonzero polynomials. Then the number of zeros of f in $(\mathbf{Z}/(m))^n$ satisfies

$$N(f) \leq m^{n-1} \prod_{i=1}^{t} d_i.$$

5.4 Computing Fixed Points

Let $f(x)$ be a mapping from $\mathbf{Z}/(m)$ to $\mathbf{Z}/(m)$. An element $x \in \mathbf{Z}/(m)$ is called a *fixed point* of f if $f(x) = x$. Thus, x is a fixed point of f if and only if it is a zero of $f(x) - x$. By the results of Section 5.1 the CRT can be used to compute the number of fixed points of a function from $\mathbf{Z}/(m)$ to $\mathbf{Z}/(m)$, and the CRA can be used to construct all the fixed points of a function f based on those of the functions f_i over $\mathbf{Z}/(m_i)$, where $f_i = f \bmod p_i$. In this section we take the RSA permutations as examples of the application of the CRT and CRA in computing the number of fixed points. Although this subject matter belongs also to Chapter 7 below, we prefer to present the material here because of its connection with fixed points. To this end, we give a brief description of the RSA public-key cryptosystem [71, 77, Chapter 4].

The RSA public-key cryptosystem supports both secrecy and authentication, and hence can provide complete and self-contained support for public-key distribution and signatures. In this system a user chooses two distinct primes p and q and computes $n = p \times q$ and $\Phi(n) = (p-1)(q-1)$. He or she then chooses e to be an integer in $[1, n-1]$ such that $\gcd(e, \Phi(n)) = 1$. Furthermore, the user finds the integer d such that $e \times d = 1 \bmod \Phi(n)$. The public parameters are n and e, while d, p, q, and $\Phi(n)$ are kept secret.

Based on these parameters the public and private transformations are respectively defined by

$$E(M) = M^e \bmod n, \quad D(C) = C^d \bmod n,$$

where $M \in [0, n-1]$ denotes the message, and $C \in [0, n-1]$ the signed message or enciphered message. Clearly, D and E are inverses. Since d is private, so is D; and since n and e are public, so is E. This constitutes a cryptosystem that can be used for both secrecy and authentication. That is, for secrecy, A sends $E_B(M)$ to B as usual; for authentication, A sends $D_A(M)$ as usual. For both secrecy and authentication, suppose first that message digests are not employed. Assuming $n_A \leq n_B$, A computes $C = E_B(D_A(M))$ and sends C to B. Then B recovers M as usual by $M = E_A(D_B(E_B(D_A(M))))$. In the case that $n_A \geq n_B$, A can instead transmit $C' = D_A(E_B(M))$. Then B can recover M as $M = D_B(E_A(D_A(E_B(M))))$.

The iteration attack on the system works as follows: If the message M is sent as $E(M) = M^e \bmod n$, let m be the order of e modulo $\Phi(n)$. Then applying E successively m times gives $M^{e^m} = M \bmod n$. In this way the message is recovered by employing only public information E. For any message M with $\gcd(M, n) = 1$, less than m iterations may be enough. Now the problem is whether such an attack is computationally feasible. This depends on the smallness of m and the factors of m. To prevent this attack, Rivest [72] suggested that p and q be chosen as follows: $p = ap' + 1$, $p' = bp'' + 1$, and $q = cq' + 1$, $q' = dq'' + 1$, where p', p'', q' and q'' are distinct primes and a, b, c and d are small integers.

The above smallest m is the integer such that x^{e^m} is equal to x, or equally every element of $\mathbf{Z}/(pq)$ is a fixed point of x^{e^m}. Thus, computing the number of fixed points of the permutations x^{e^i} is an important issue of the RSA system. This can be done by the bridge supported by the CRT in Section 5.1.

Let $F_i(x) = x^{e^i} - x$, then x is a fixed point of $E^i(x)$ if and only if x is a zero of $F_i(x)$, where $E(x) = x^e$ and $E^i = E(E^{i-1})$.

Theorem 5.4.1 *Let the symbols be the same as before. Then*

$$N(F_1) = [1 + \gcd(e - 1, p - 1)][1 + \gcd(e - 1, q - 1)]. \tag{5.4}$$

Proof: Let $e_1 = e \bmod (p - 1)$, $e_2 = e \bmod (q - 1)$ and

$$f_p(x) = F_1(x) \bmod p = x^{e_1} - x, \quad f_q(x) = F_1(x) \bmod q = x^{e_2} - x.$$

We now compute $N(F_1)$. Let α be a primitive root modulo p. Then every nonzero element of $Z/(p)$ can be written as α^j for some integer j with $0 \le j \le p - 2$. Obviously, $f_p(x) = 0$ if and only if $x = 0$ or $x^{e_1 - 1} = 1$. To compute the zeros of $x^{e_1 - 1} - 1$ over $Z/(p)$, let $x = \alpha^j$. Then $x^{e_1 - 1} = \alpha^{j(e_1 - 1)} = 1$ if and only if there is an integer k such that $j(e_1 - 1) = k(p - 1)$. Let $g = \gcd(e_1 - 1, p - 1)$. Then $j = [k(p - 1)/g]/[(e_1 - 1)/g]$ is an integer if and only if $k = k_1(e_1 - 1)/g$, which is equivalent to $j = k_1(p - 1)/g$. From $0 \le j \le p - 2$ it follows that

$$0 \le k_1 \le \frac{(p - 2)g}{p - 1} = g - \frac{g}{p - 1}.$$

Since $0 < g = \gcd(e_1 - 1, p - 1) \le e_1 - 1 \le p - 3$, we get $0 \le k_1 \le g - 1$. Thus, $N(f_p) = 1 + g = 1 + \gcd(e_1 - 1, p - 1)$. By definition $e_1 = e \bmod (p - 1)$. It then follows easily that $\gcd(e_1 - 1, p - 1) = \gcd(e - 1, p - 1)$. Thus,

$$N(f_p) = 1 + g = 1 + \gcd(e - 1, p - 1).$$

Similarly, we have $N(f_q) = 1 + \gcd(e - 1, q - 1)$. Thus the equation (5.4) follows easily from Theorem 5.1.1.   □

Generally we have the following conclusion which follows from (5.4).

Theorem 5.4.2 *Let the symbols be the same as before. Then*

$$N(F_i) = \left[1 + \gcd(e^i - 1, p - 1)\right]\left[1 + \gcd(e^i - 1, q - 1)\right].$$

As a corollary we have the following conclusion.

Theorem 5.4.3 *Let the symbols be the same as before. Then $N(F_{2^h})$ is equal to*

$$\left[1 + \gcd((e - 1) \prod_{i=0}^{2^h - 1} (e^{2^i} - 1), p - 1)\right] \times$$

$$\left[1 + \gcd((e - 1) \prod_{i=0}^{2^h - 1} (e^{2^i} + 1), q - 1)\right].$$

If one of the conditions $e = kl + 1$ and $e^{2^i} = kl - 1$ for some k and for some i with $0 \leq i \leq h - 1$ holds, where $l = \mathrm{lcm}\{p-1, q-1\}$, then $N(F_{2^h}) = pq$ and thus every element of $\mathbf{Z}/(n)$ is a fixed point of the RSA permutation $E^{2^h}(x) = x^{e^{2^h}}$.

It follows from this theorem that the value $l = \gcd\{p - 1, q - 1\}$ should be quite large. In addition, the exponent e should be chosen such that $e^{2^i} + 1$ is not divisible by l for i between 1 and i_0, where i_0 should be at least 40 considering the present computing power.

Assume that the primes p are q are chosen such that $p = 2p_1 + 1$ and $q = 2q_1 + 1$, where p_1 and q_1 are also odd primes, such p and q are often called Sophie German primes. Then e should be chosen such that the orders of e modulo p_1 and q_1 are large enough, due to the following result which follows easily from Theorem 5.4.1.

Theorem 5.4.4 *Let $p = 2p_1 + 1$ and $q = 2q_1 + 1$, where p, p_1, q, q_1 are all odd primes, and let $a = \mathrm{lcm}\{ord_{p_1}(e), ord_{q_1}(e)\}$. Then*

$$N(F_a) = pq.$$

The following result follows also from Theorem 5.4.1, and explains why the primes p_1 and q_1 should be chosen such that $p_1 - 1$ and $q_1 - 1$ have also large prime factors.

Theorem 5.4.5 *Let $p = 2^k p_1 + 1$ and $q = 2^l q_1 + 1$, where p, p_1, q, q_1 are all odd primes, and let $b = \mathrm{lcm}\{ord_{p_1}(e), ord_{q_1}(e)\}$. Then*

$$N(F_b) \geq [1 + 2p_1][1 + 2q_1].$$

If $e^b - 1 = 2^r e_1$, where e_1 is odd, then

$$N(F_b) = [1 + 2^{\min\{r,k\}} p_1][1 + 2^{\min\{r,l\}} q_1].$$

Generally, let $f(x)$ be a polynomial over $\mathbf{Z}/(m)$, where $m = m_1 m_2 \cdots m_t$ and m_i are pairwise relatively prime, and let $F(x) = f(x) - x$, $F_i(x) = F(x) \bmod m_i$. Then the number of fixed points of $f(x)$ is given by

$$N(F) = N(F_1) N(F_2) \cdots N(F_t).$$

Thus, to compute the number of fixed points of $f(x)$, we need only to compute the number of zeros of $F_i(x)$. If exact fixed points are our concern, we may first compute the zeros of $F_i(x)$, then get all the exact fixed points of $f(x)$ by the CRA.

5.5   Bridging Divisions of Polynomials

Throughout this section p_i are distinct primes, and $m = p_1 p_2 \cdots p_t$. The polynomials concerned are formal polynomials. This means that two polynomials are considered equal if and only if their corresponding coefficients are equal. As usual, the expression $f(a)$ means the value of $f(x)$ when the latter is regarded as a polynomial function. For two nonzero polynomials $f(x), g(x) \in F[x]$, where F is a field, there are two polynomials $q(x), r(x) \in F[x]$ such that

$$f(x) = g(x)q(x) + r(x), \tag{5.5}$$

where $r(x) = 0$ or $\deg(r) < \deg(g)$. This means that $F[x]$ is an Euclidean domain. This is however not true for polynomials of $\mathbf{Z}/(m)[x]$, since some elements of $\mathbf{Z}/(m)[x]$ have zero divisors.

Let $f(x), g(x) \in \mathbf{Z}/(m)[x]$. We say that $f(x)$ is divisible by $g(x)$ if there is a polynomial $h(x) \in \mathbf{Z}/(m)[x]$ such that $f(x) = g(x)h(x)$. We then write $g(x)|f(x)$. The equation $f(x) = g(x)h(x)$ may be satisfied by several polynomials $h(x)$. Let

$$G = \{d(x) \in \mathbf{Z}/(m)[x] : d(x)g(x) = 0\}.$$

It is easily seen that G is an Abelian group with respect to the polynomial addition of $\mathbf{Z}/(m)[x]$. If h is one polynomial such that $f(x) = g(x)h(x)$, then all the polynomials of $h(x) + G$ are such polynomials, where $h(x) + G$ denotes all polynomials $h(x) + g'(x)$ with $g'(x) \in G$. On the other hand, if $h_1(x)$ satisfies $f(x) = g(x)h_1(x)$, then $g(x)(h(x) - h_1(x)) = 0$. It follows that $h_1(x) \in h(x) + G$. Thus, finding one $h(x)$ means that all such $h(x)$ are determined if the set G is known.

Consider now the two polynomials $f(x) = 4x^2$ and $g(x) = 2x$ over $\mathbf{Z}/(6)$. One $h(x)$ such that $f(x) = g(x)h(x)$ is $h(x) = 2x$. It is easily seen that the set G corresponding to $g(x) = 2x$ is

$$G = \left\{ d(x) = \sum_i d_i x^i : d_i = 0 \bmod 3 \right\}.$$

Thus, there are infinitely many $h(x) \in \mathbf{Z}/(6)[x]$ such that $f(x) = g(x)h(x)$.

Now the problem is how to find one $h(x)$ such that $f(x) = g(x)h(x)$, given $f(x), g(x) \in \mathbf{Z}/(m)[x]$. This can be done by the CRT and CRA. Let $f_i(x) = f(x) \bmod p_i$ and $g_i(x) = g(x) \bmod p_i$ for $i = 1, ..., t$. If $g_i(x) = 0$, then $f_i(x)$ must be equal to zero. In this case any $h_i(x) \in \mathbf{Z}/(p_i)[x]$ satisfies $f_i(x) = g_i(x)h_i(x)$. If $g_i(x) \neq 0$, then in the field $\mathbf{Z}/(p_i)$ we solve the equation $f_i(x) = g_i(x)h_i(x)$. In this way we can obtain t polynomials

$$h_i(x) = \sum_{j=0}^{l_i} h_{i,j} x^j \in \mathbf{Z}/(p_i)[x].$$

Let $M_i = m/p_i$, and $l = \max\{l_1, ..., l_t\}$. With Euclidean algorithm we compute two integers u_i and v_i such that $u_i M_i + v_i p_i = 1$. Let $h_{i,j} = 0$ for $j = l_i + 1, ..., l$ and

$$h_j = \sum_{i=1}^{t} h_{i,j} u_i M_i \bmod m.$$

Then the polynomial $h(x) = \sum_{i=0}^{l} h_i x^i$ satisfies $f(x) = g(x)h(x)$.

It is obvious that for three polynomials $f(x), g(x)$ and $h(x)$, the equation $f(x) = g(x)h(x)$ does not imply that $\deg(f) \geq \max\{\deg(g), \deg(h)\}$. Another problem is how to find polynomials $h(x)$ and $r(x)$ such that $f(x) = g(x)h(x) + r(x)$, given two polynomials $f(x), g(x) \in \mathbf{Z}/(m)[x]$. Let the set G be as before. If $h_0(x) \in \mathbf{Z}/(m)[x]$ is a polynomial such that (5.5) holds, then all such $h(x)$ are given by the set $h_0(x) + G$. We have a similar algorithm to solve this problem.

A generalization of the polynomial division problem is the solving of a set of linear equations

$$f_{i,1} g_1 + f_{i,2} g_2 + \cdots + f_{i,n} g_n = h_i, \quad i = 1, 2, \cdots, k,$$

where $f_{i,j}, h_i \in \mathbf{Z}/(m)[x]$ are given polynomials and g_i are unknown. We can similarly transform this problem to problems over $\mathbf{Z}/(p_i)$ and use the CRA to get all the solutions.

5.6   Permutation Polynomials of $\mathbf{Z}/(m)$

A polynomial $\pi(x) \in GF(q)[x]$ (resp. $\mathbf{Z}/(m)[x]$) is called a *permutation polynomial* of $GF(q)$ (resp. $\mathbf{Z}/(m)$) if it is a one-to-one mapping from $GF(q)$ (resp. $\mathbf{Z}/(m)$) to $GF(q)$ (resp. $\mathbf{Z}/(m)$) when it is considered as a polynomial function.

Both the encryption and decryption transformation of most cryptosystems are permutations. The permutation polynomials x^e of $\mathbf{Z}/(pq)$ form both the encryption and decryption transformation of the RSA public-key cryptosystem [71, 77]. Certain permutation polynomials of $\mathbf{Z}/(m)$ can be used to construct sequence generators as follows. Let m be a positive integer and $\pi(x)$ be a permutation polynomial of $\mathbf{Z}/(m)$. A sequence based on the permutation polynomial is given by

$$s_i = \pi(i) \bmod 2, \quad i = 0, 1, 2, ...$$

The "pattern distributions" of the sequence are determined by the "nonlinearity" of the permutation polynomial π. On the other hand, the permutation polynomial should be of some special form which has a fast implementation. In this sections we use the CRT and CRA to bridge the permutation polynomials over $\mathbf{Z}/(m)$ and those over $\mathbf{Z}/(m_i)$, where $m = m_1 m_2 \cdots m_t$ and m_i are pairwise relatively prime.

Any mapping from $GF(q)$ to $GF(q)$ can be represented as a polynomial of $GF(q)[x]$, but this is not true for $\mathbf{Z}/(m)$, where m is composite. In this section we construct all permutation polynomials of $\mathbf{Z}/(p_1 \cdots p_t)$, where p_i are pairwise distinct primes, and concentrate on those having a fast implementation. Also we will construct a class of permutation polynomials for any ring $\mathbf{Z}/(m)$. The bridge for transforming some permutation polynomials of finite fields to those of $\mathbf{Z}/(m)$ is again the CRT.

The following theorem is a bridge between permutation polynomials of $\mathbf{Z}/(m)$ and those of $\mathbf{Z}/(m_i)$ for $i = 1, 2, ..., t$, where $m = m_1 m_2 \cdots m_t$ and $m_1, m_2, ..., m_t$ are pairwise relatively prime [67].

Theorem 5.6.1 *Let $m_1, ..., m_t$ be pairwise relatively prime and $m = \prod_{i=1}^{t} m_i$. If*

$$\pi(x) = h_0 + h_1 x + \cdots + h_l x^l \in \mathbf{Z}/(m)[x] \qquad (5.6)$$

is a permutation polynomial of $\mathbf{Z}/(m)$, then for each i, where $1 \leq i \leq t$, the polynomial

$$\pi_i(x) = h_{i,0} + h_{i,1} x + \cdots + h_{i,l} x^l \in \mathbf{Z}/(m)[x] \qquad (5.7)$$

is a permutation polynomial of $\mathbf{Z}/(m_i)$, where $h_{i,j} = h_j \bmod m_i$ for $0 \leq j \leq l$.

Conversely, if $\pi_i(x)$ is a permutation polynomial of $\mathbf{Z}/(m_i)$ for $i = 1, ..., t$, where some of the higher-order coefficients $h_{i,l}, h_{i,l-1}, ...$ could be zero, then the unique polynomial $\pi(x)$ satisfying

$$\pi(x) \bmod m_i = \pi_i(x), \quad i = 1, 2, ..., t \qquad (5.8)$$

is a permutation of $\mathbf{Z}/(m)$.

Proof: Let $\phi(y) = (y \bmod m_1, y \bmod m_2, ..., y \bmod m_t)$, where $y \in \mathbf{Z}/(m)$. The CRT says that ϕ is a ring isomorphism between $\mathbf{Z}/(m)$ and $\mathbf{Z}/(m_1) \times \mathbf{Z}/(m_2) \times \cdots \times \mathbf{Z}/(m_t)$. Clearly,

$$\phi(\pi(x)) = (\pi_1(x \bmod m_1), \pi_2(x \bmod m_2), ..., \pi_t(x \bmod m_t)). \qquad (5.9)$$

If $\pi_1(x_1) = \pi_1(x_2)$ for two distinct elements $x_1, x_2 \in \mathbf{Z}/(m_1)$, then the two elements $\phi^{-1}(x_1, 0, ..., 0)$ and $\phi^{-1}(x_2, 0, ..., 0)$ of $\mathbf{Z}/(m)$ are distinct, but still have the same image under the bijection $\phi\pi$, where $\phi\pi$ is the composition of ϕ and π. This is a contradiction. Thus, $\pi_1(x)$ is a permutation of $\mathbf{Z}/(m_1)$. The same conclusion can be proved for other $\pi_i(x)$.

The CRT guarantees the existence and uniqueness of the $\pi(x)$ of Equation (5.8). If $\pi(x) = \pi(y)$, where $x \neq y$, then $(\pi_1(x \bmod m_1), ..., \pi_t(x \bmod m_t)) = (\pi_1(y \bmod m_1), ..., \pi_t(y \bmod m_t))$. Since $x \neq y$, then $(x \bmod m_1, ..., x \bmod m_t) \neq (y \bmod m_1, ..., y \bmod m_t)$. It follows that there must exist at least one i such that $x \bmod m_i \neq y \bmod m_i$. Thus, $\pi_i(x)$ is not a permutation of $Z/(m_i)$. This is a contradiction, and proves part two. □

Consider permutation polynomials $\pi_i(x) = h_{i,0} + h_{i,1}x + \cdots + h_{i,l}x^l \in Z/(m_i)[x]$ for all $i = 1, ..., t$, where some of the higher order coefficients may be zero. The polynomial given by (5.8) is calculated by the CRA as follows. Let $M_i = m/m_i$. Since M_i and m_i are relatively prime, with the Euclidean algorithm we can compute two integers u_i and v_i such that $M_i u_i + m_i v_i = 1$. Put

$$h_j = \sum_{i=1}^{t} h_{i,j} u_i M_i \bmod m. \tag{5.10}$$

Then $\pi(x) = h_0 + h_1 x + \cdots + h_l x^l$ is the permutation polynomial defined by (5.8). In what follows we shall construct permutations on $Z/(m)$ from those on $Z/(m_i)$ by making use of the CRA and Theorem 5.6.1.

One important class of permutation polynomials on $Z/(m)$, where $m = p_1 p_2 \cdots p_t$ and p_i are pairwise distinct primes, is described by the following theorem. The interconnection with the RSA public-key cryptosystem is obvious.

Theorem 5.6.2 *Let $m = p_1 p_2 \cdots p_t$, where p_i are pairwise distinct primes, and let e_i be a positive integer such that $\gcd(e_i, (p_i - 1)) = 1$ for $i = 1, 2, ..., t$. Then*

1. *$\pi(x) = \sum_{i=1}^{t}(m/p_i)w_i x^{e_i}$ is a permutation of $Z/(m)$ if $\gcd(p_i, w_i) = 1$ for $i = 1, 2, ..., t$, where w_i are integers.*

2. *The inverse permutation of $\pi(x)$ is $\pi'(x) = \sum_{i=1}^{t}(m/p_i)u_i x^{d_i}$, where $u_i = c_i b_i^{d_i} \bmod p_i$, c_i is the multiplicative inverse of m/p_i modulo p_i, b_i is the multiplicative inverse of a_i modulo p_i, $a_i = w_i m/p_i \bmod p_i$, and d_i is the unique integer such that $1 \leq d_i \leq p_i - 1$ and $e_i d_i = 1 \bmod (p_i - 1)$.*

Proof: $\pi(x) \bmod p_i = (mw_i/p_i \bmod p_i)x^{e_i}$ is a permutation of $Z/(p_i)$, since $mw_i/p_i \neq 0 \bmod p_i$ and $\gcd(e_i, p_i - 1) = 1$. The conclusion of part one then follows from Theorem 5.6.1.

By definition and assumptions $\pi_i(x) = \pi(x) \bmod p_i = a_i x^{e_i}$ and $\pi'_i(x) = \pi'(x) \bmod p_i = b_i^{d_i} x^{d_i}$. Then $\pi_i(\pi'_i(x)) = a_i b_i^{e_i d_i} x^{e_i d_i} = a_i b_i x = x$. Similarly, we have $\pi'_i(\pi_i(x)) = x$. Thus, $\pi'_i(x)$ is the inverse of $\pi_i(x)$. By the CRT $\pi'(x)$ is the inverse of $\pi(x)$. □

This theorem shows that this class of permutation polynomials is invariant under the inverse operation. By choosing w_i and e_i we get $(p_i - 1)\Phi(p_i - 1)$

permutations of the form $(w_i m/p_i)x^{e_i}$ of $\mathbf{Z}/(p_i)$. Thus, by choosing the w_i and e_i in Theorem 5.6.2, we get altogether $\prod_{i=1}^{t}(p_i-1)\Phi(p_i-1)$ distinct permutation polynomials of such a form on $\mathbf{Z}/(m)$.

When choosing $w_1 = w_2 = \cdots = w_t = 1$ in Theorem 5.6.2, we obtain the following theorem.

Theorem 5.6.3 *Let* $m = p_1 p_2 \cdots p_t$, *where* p_i *are pairwise distinct primes, and let* e_i *be a positive integer such that* $\gcd(e_i, (p_i - 1)) = 1$ *for* $i = 1, 2, ..., t$. *Then* $\sum_{i=1}^{t}(m/p_i)x^{e_i}$ *is a permutation of* $\mathbf{Z}/(m)$.

By choosing w_i such that $w_i m/p_i = 1 \bmod p_i$ and $1 \leq w_i \leq p_i - 1$ for each i in Theorem 5.6.2, we obtain the following theorem.

Corollary 5.6.4 *Let* $m = p_1 p_2 \cdots p_t$, *where* p_i *are pairwise distinct primes, and let* e *be a positive integer such that* $\gcd(e, (p_1 - 1)(p_2 - 1) \cdots (p_t - 1)) = 1$. *Then* x^e *is a permutation of* $\mathbf{Z}/(m)$.

This corollary was proved by Cordes in 1976 with a direct proof [24], two years before the discovery of the RSA public-key cryptosystem [77, Chapter 4], where a permutation of this form for the case $t = 2$ is used.

In practical applications a fast implementation of the permutation x^e of Corollary 5.6.4 is necessary. A fast software algorithm is the following. To compute x^e modulo m, where m is in some applications quite large, we first compute $e_i = e \bmod (p_i - 1)$. Then for i from 1 to t, we compute $y_i = (x \bmod p_i)^{e_i}$. Then we use the CRA given in (5.10) to compute the unique y such that $y \bmod p_i = y_i$ for $i = 1, ..., t$, where $0 \leq y \leq m - 1$. Then $y = x^e$ is the one we want to obtain. The computation of y_i can be done by any fast exponentiation algorithm. If $e \geq \max\{p_1, p_2, ..., p_t\}$. This modular algorithm must be more efficient than the fast exponentiation algorithm for the direct computation of x^e.

In what precedes we have constructed a number of permutation polynomials of $\mathbf{Z}/(p_1 \cdots p_t)$ based on the special permutations x^{e_i} of $\mathbf{Z}/(p_i)$. To get more permutations of $\mathbf{Z}/(m)$, we need permutations of forms other than x^e on $\mathbf{Z}/(p_i)$. It is known that $x^{(p+1)/2} + ax \in Z/(p)[x]$ is a permutation of $Z/(p)$, where p is an odd prime, if $a = (c^2 + 1)/(c^2 - 1)^{-1}$ for some $c \in Z/(p)^*$ with $c^2 \neq 1$ [50, Theorem 7.13].

Theorem 5.6.5 *Let* $m = p_1 p_2 \cdots p_t$, *where* p_i *are pairwise distinct odd primes, and let* $a_i = (c_i^2 + 1)/(c_i^2 - 1)^{-1}$ *for some* $c_i \in \mathbf{Z}/(p_i)$ *with* $c_i^2 \neq \pm 1$, *where* $1 \leq i \leq t$. *Then*

$$\pi(x) = \sum_{i=1}^{t}(m/p_i)w_i x^{(p_i+1)/2} + \left(\sum_{i=1}^{t} a_i m w_i/p_i\right) x$$

is a permutation of $\mathbf{Z}/(m)$ *if* $\gcd(p_i, w_i) = 1$ *for* $i = 1, 2, ..., t$.

Proof: Since $x^{(p_i+1)/2} + a_i x$ is a permutation polynomial of $\mathbf{Z}/(p_i)$ and $m w_i / p_i \bmod p_i \neq 0$, the polynomial $(m w_i / p_i \bmod p_i)[x^{(p_i+1)/2} + a_i x]$ is a permutation of $\mathbf{Z}/(p_i)$. By definition $\pi(x) \bmod p_i = (m w_i / p_i \bmod p_i)[x^{(p_i+1)/2} + a_i x]$, it then follows from Theorem 5.6.1 that $\pi(x)$ is a permutation of $\mathbf{Z}/(m)$.   □

In $\mathbf{Z}/(p)$, where p is an odd prime, it is seen that $(c_1^2 + 1)/(c_1^2 - 1)^{-1} = (c_2^2 + 1)/(c_2^2 - 1)^{-1}$ if and only if $c_1 = \pm c_2$. Thus, there are $(p - 3)/2$ distinct permutation polynomials of the form $x^{(p+1)/2} + ax$ on $\mathbf{Z}/(p)$, and $(p-1)(p-3)/2$ distinct permutation polynomials of the form $w_i(x^{(p+1)/2} + ax)$. Thus, altogether there are $\prod_{i=1}^{t}(p_i - 1)(p_i - 3)/2$ distinct permutation polynomials in this class.

It is known that $x(x^{(p-1)/2} - a)^2$ is a permutation of $Z/(p)$ if and only if $a \neq \pm 1$, where p is an odd prime [50, p.389]. Similar to Theorem 5.6.5, one can prove the following conclusion.

Theorem 5.6.6 *Let* $m = p_1 p_2 \cdots p_t$, *where* p_i *are pairwise distinct odd primes, and let* $a_i \in \mathbf{Z}/(p_i)$ *with* $a_i \neq \pm 1$, *where* $1 \leq i \leq t$. *Then*

$$\pi(x) = \sum_{i=1}^{t}(m/p_i) w_i (x^{(p_i-1)/2} - a_i)^2 x$$

is a permutation of $\mathbf{Z}/(m)$ *if* $\gcd(p_i, w_i) = 1$ *for* $i = 1, 2, ..., t$, *where* w_i *are integers.*

If $a_i = 0$, then $x(x^{(p-1)/2} - a_i)^2 = x$ is linear and is not interesting to us. By choosing w_i and $a_i \neq 0$, we get $\prod_{i=1}^{t}(p_i - 1)(p_i - 3)$ distinct permutation polynomials of $\mathbf{Z}/(m)$ in this class.

In what precedes we have presented some permutation polynomials of special forms. All permutations of $\mathbf{Z}/(m)$, where $m = p_1 p_2 \cdots p_t$, can be constructed as in the following theorem.

Theorem 5.6.7 *There are altogether* $p_1! p_2! \cdots p_t!$ *permutation polynomials of* $\mathbf{Z}/(m)$, *and each of them can be expressed as*

$$\pi(x) = \sum_{i=1}^{t}(m/p_i) u_i \sum_{k=0}^{p_i-1}(p_i - b_{i,k}) \prod_{j=0, \, j \neq k}^{p_i-1}(x + p_i - j) \bmod m, \qquad (5.11)$$

where u_i *is the integer such that* $(m/p_i) u_i + p_i v_i = 1$, *and* $b_{i,j} \in \mathbf{Z}/(p_i)$ *such that* $\{b_{i,j} : j = 0, 1, ..., p_i - 1\} = \mathbf{Z}/(p_i)$.

Proof: Since each permutation of $\mathbf{Z}/(p_i)$ can be expressed as a polynomial, there are $p_i!$ permutation polynomials of $\mathbf{Z}/(p_i)$. It then follows from the CRT that there are altogether $p_1! p_2! \cdots p_t!$ distinct permutation polynomials of $\mathbf{Z}/(m)$.

Let $\pi(x)$ be a permutation polynomial of $\mathbf{Z}/(m)$, and $\pi_i(x) = \pi(x) \bmod p_i$ for $i = 1, 2, ..., t$. Then $\pi_i(x)$ is a permutation polynomial of $\mathbf{Z}/(p_i)$. Assume that $\pi_i(j) = b_{i,j}$, then $\{b_{i,j} : j = 0, 1, ..., p_i - 1\} = \mathbf{Z}/(p_i)$ since π_i is a permutation. Note that $\pi_i(j) = b_{i,j}$ if and only if $\pi_i(x) = b_{i,j} \bmod (x - j)$. Let $n_j(x) = (x - j)$, where $j \in \mathbf{Z}/(p_i)$ and

$$N(x) = \prod_{j=0}^{p_i-1} (x - j), \quad N_j(x) = N(x)/n_j(x).$$

Since $0, 1, ..., p_i - 1$ are pairwise distinct, the polynomials $(x - j)$ are pairwise relatively prime. By definition the polynomials $N_j(x)$ and $n_j(x)$ are pairwise relatively prime. With Euclidean algorithm we compute two polynomials $s_j(x)$ and $t_j(x)$ such that $N_j(x)s_j(x) + n_j(x)t_j(x) = 1$, from which it follows that $s_j(x) = N_j(j)^{-1} \bmod n_j(x)$. Thus, the CRA for polynomials gives

$$\pi_i(x) = [\sum_{h=0}^{p_i-1} b_{i,h}s_h(x)N_h(x) \bmod N(x)] = \sum_{h=0}^{p_i-1} b_{i,h}s_h(x)N_h(x)$$

$$= \sum_{h=0}^{p_i-1} b_{i,h} \frac{N_h(x)}{N_h(h)} = \sum_{h=0}^{p_i-1} b_{i,h}L_h(x),$$

where $L_h(x) = \prod_{j \neq h}(x - j)/(h - j) \in \mathbf{Z}/(p_i)[x]$. By Wilson's theorem $\prod_{j \neq h}(h - j) = (p_i - 1)! = -1 \bmod p_i$. Hence

$$\pi_i(x) = \sum_{k=0}^{p_i-1} -b_{i,k} \prod_{j \neq k} (x - j)$$

$$= \sum_{k=0}^{p_i-1} (p_i - b_{i,k}) \prod_{j \neq k} (x + p_i - j).$$

Then by applying the CRA for integers we get the expression of $\pi(x)$ in (5.11). This proves the theorem. □

   In cryptography we are much interested in permutation polynomials of $\mathbf{Z}/(p^2)$ with ideal nonlinearity and a fast implementation. It is easily seen that there is no permutation polynomial of the form x^e on $\mathbf{Z}/(p^2)$ with $e \geq 2$. But it clear that x^e is a permutation of $(\mathbf{Z}/(p^2))^*$ if and only if $\gcd(e, p(p-1)) = 1$ since $(\mathbf{Z}/(p^2))^*$ is a multiplicative group of order $p(p-1)$. On the other hand, $-x$ is a permutation of the set $R = \{0, p, 2p, ..., (p-1)p\}$. Thus, we can combine the two permutations together, in order to get a permutation of $\mathbf{Z}/(p^2)$. Generally, we have the following result.

Theorem 5.6.8 *Let p be a prime, k and e be positive integers such that $\gcd(e, p(p-1)) = 1$ and $k \leq \min\{e, p^{k-1}(p-1)\}$. Then*

$$\pi(x) = (x^{p^{k-1}(p-1)} - 1)x + x^e$$

is a permutation of $\mathbf{Z}/(p^k)$.

Proof: Since $(\mathbf{Z}/(p^k))^*$ is a group of order $p^{k-1}(p-1)$ and $k \leq \min\{e, p^{k-1}(p-1)\}$,

$$\pi(x) = \begin{cases} -x, & x \in R, \\ x^e, & x \in (\mathbf{Z}/(p^k))^*, \end{cases}$$

where $R = \{0, p, 2p, ..., (p-1)p\}$. Thus, π maps R into R, and $(\mathbf{Z}/(p^k))^*$ into $(\mathbf{Z}/(p^k))^*$. Obviously, π is a permutation of R. Since $\gcd(e, p(p-1)) = 1$ and the order of $(\mathbf{Z}/(p^k))^*$ is $p^{k-1}(p-1)$, $\pi(x)$ is a permutation of $\mathbf{Z}/(p^k)^*$. This proves the theorem.   □

With the help of Theorems 5.6.2 and 5.6.8, one can similarly prove the following conclusion.

Theorem 5.6.9 *Let $m = p_1^{k_1} \cdots p_t^{k_t}$, where p_i are pairwise distinct primes and $k_i \geq 1$ are integers. For each i with $1 \leq i \leq t$, choose a positive integer e_i such that $k_i \leq \min\{e_i, p_i^{k_i-1}(p_i - 1)\}$ and $\gcd(e_i, p_i^{k_i-1}(p_i - 1)) = 1$. Then*

$$\pi(x) = \sum_{i=1}^{t} (m/p_i^{k_i}) w_i [(x^{p_i^{k_i-1}(p_i-1)} - 1)x + x^{e_i}]$$

is a permutation polynomial of $\mathbf{Z}/(m)$ if and only if $\gcd(w_i, p_i) = 1$ for each i.

Based on the permutation of Theorem 5.6.8, $p_1^{k_1-1}(p_1 - 1) \cdots p_t^{k_t-1}(p_t - 1)$ permutations of $\mathbf{Z}/(m)$ are described by Theorem 5.6.9, where $m = p_1^{k_1} \cdots p_t^{k_t}$. In fact if, for some i, $k_i = 1$, further permutation polynomials of $\mathbf{Z}/(m)$ can be constructed by the permutation of $\mathbf{Z}/(p^k)$ for $k \geq 2$ and those of $Z/(p)$ used in Theorem 5.6.6. Thus, in this section all permutation polynomials of any $\mathbf{Z}/(m)$ were constructed, where m has no square factors. A class of permutation polynomials over $\mathbf{Z}/(m)$ were also constructed, where m has square factors.

If we can get more permutations of the rings $\mathbf{Z}/(p^e)$, more permutations of $\mathbf{Z}/(m)$ will be constructed with the CRA and CRT. The following conclusion due to Nöbauer should be helpful and the proof is easy [67].

Theorem 5.6.10 *A polynomial $f(x) = f_0 + f_1 x + \cdots + f_n x^n \in \mathbf{Z}[x]$ is a permutation polynomial of $\mathbf{Z}/(p^2)$ if and only if $f(x) \bmod p$ is a permutation polynomial of $\mathbf{Z}/(p)$ and the congruence $f'(x) = 0 \bmod p$ has no solutions, where $f'(x) = f_1 + 2f_2 x + 3f_3 x^2 + \cdots n f_n x^{n-1}$.*

If $f(x) \in \mathbf{Z}[x]$ is a permutation polynomial of $\mathbf{Z}/(p^2)$, then it is also a permutation polynomial of $\mathbf{Z}/(p^k)$ for all positive integers $k \geq 1$.

Chapter 6

In Coding Theory

Coding theory has important applications in communications and computer systems, where data errors frequently occur. The basic idea for error detecting and correcting is to add redundancy to the data which will be sent via a noisy channel or stored in a computer. Remainder techniques based on the CRT are quite useful in constructing error-correcting codes. We shall distinguish between two kinds of applications of the CRT in error-correcting. When applying the CRT for the Euclidean domain $GF(q)[x]$, we are concerned with redundant residue codes over finite fields; while considering the CRT for the Euclidean domain $\mathbf{Z}$ we will investigate the arithmetic codes. Thus, in this chapter we are concerned with the applications of CRT in block coding and arithmetic coding.

This chapter first gives a detailed treatment of redundant residue codes based on the CRT. Known results will be summarized, and new codes will be constructed and analyzed. As we shall see, redundant residue codes and generalized redundant residue codes include many known good codes, and the family of subfield codes of the generalized redundant residue codes contains more good block codes. It is interesting to note that all of the encoding and decoding as well as the parameter analysis of the generalized redundant residue codes can be carried out with the CRT and CRA.

Then this chapter introduces the application of the CRT for the Euclidean domain $\mathbf{Z}$ in constructing arithmetic redundant residue codes. The motivation of constructing arithmetic residue codes is the error-detection and -correction in residue number systems [6, 7, 53, 54, 56]. Redundant residue codes constitute an interesting topic.

There are quite a number of open problems in this field which need to be investigated. For instance, can every linear code over a finite field be viewed as a redundant residue code? If not, which linear codes are redundant residue codes? How to construct good redundant residue codes? Is there any efficient decoding algorithm for good redundant residue codes?

6.1   Basics of Block Codes

Error-correcting codes are divided into two classes: block and convolutional codes. For block codes the encoding (resp. decoding) device has no internal memory, while for convolutional codes it has. This is the clear distinction between
these two classes of codes. Since block codes over finite fields are relevant to
our applications in this chapter, we describe some basic notions and notations
of block codes in this section.

An (n, M, d) code C over an alphabet A is a set of M vectors of length n
with components from A such that any two vectors differ in at least d places,
and d is the maximum number with this property. Each vector of C is called
a *codeword*. The parameter d is called the *minimum distance* of the code. The
Hamming distance of two codewords is the number of components in which the
two codewords differ. To illustrate these notions, we take the *repetition code* as
an example. Let $A = \{0, 1\}$, then the set of vectors $C = \{00000, 11111\}$ is a
$(5, 2, 5)$ code over A. The distance between the two codewords 00000 and 11111
is thus 5. With this code we can encode a binary message by replacing each
0 (resp. 1) of the message with 00000 (resp. 11111). Suppose the message is
100011, then the encoded message is

$$1111100000000000000001111111111.$$

If an encoded message is sent via a noisy channel, some bits of the received
one may not be the same as the original ones. Let $\mathbf{r}$ be the received vector
corresponding to the original message. If we assume that in each block, $\mathbf{r}_j =
(r_{5j}r_{5j+1}\cdots r_{5j+4})$, there are at most two errors, i.e, at most two bits which are
different from the original ones, then all errors of the received vector can be
corrected. To correct possible errors, compute the distances between a block $\mathbf{r}_j$
and 00000, 11111 in turn. By assumption one and only one of the two distances
must be less than or equal to 2. Then the original bit of the message must be
1 (resp. 0) if the distance between $\mathbf{r}_j$ and 11111 (resp. 00000) is less than or
equal to 2. Thus, the above code is able to correct two random errors, and is
said to be a 2-error correcting code.

The above code is very simple in the sense that the encoding and decoding
procedure are quite easy, but very inefficient in the sense that five bits are
needed to encode a one-bit message. To construct codes which have an easy
encoding and decoding algorithm, alphabets with some algebraic structure, such
as groups, rings, or fields, are needed. Codes over finite fields are the most
popular ones.

Let F be a finite field, by definition a code C over F is a subset of the vector
space F^n. If C is a linear subspace of F^n over F, it is called a *linear code*;

otherwise it is a *nonlinear code*. The above (5, 2, 5) repetition code is a linear code over $GF(2)$, while the code $C = \{0000, 1100, 1111\}$ is not linear.

If C is a linear code of length n over F, the dimension, say k, of C must be no more than n. Such a C is called an $[n, k]$ linear code. Thus, the number of codewords in C must be q^k if the field F is a finite field of q elements. Any $k \times n$ matrix over F whose k row vectors generate the linear space C linearly, i.e., the minimum linear subspace of F^n containing these k row vectors is equal to C, is called a *generator matrix* of the code C.

For two vectors **u** and **v** of F^n, where $\mathbf{u} = (u_0, u_1, \cdots, u_{n-1})$ and $\mathbf{v} = (v_0, v_1, \cdots, v_{n-1})$, the *inner product* or *scalar product* is defined to be

$$\mathbf{u} \cdot \mathbf{v} = \sum_{i=0}^{n-1} u_i v_i.$$

If $\mathbf{u} \cdot \mathbf{v} = 0$, **u** and **v** are called *orthogonal*. If C is an $[n, k]$ linear code over F, its *dual* or *orthogonal code* $C^\perp$ is the set of vectors which are orthogonal to all codewords of C, that is,

$$C^\perp = \{\mathbf{u} | \mathbf{u} \cdot \mathbf{v} = 0 \text{ for all } \mathbf{v} \in C\}.$$

By elementary algebra the dual code $C^\perp$ is an $[n, n-k]$ linear code, and $(C^\perp)^\perp = C$. A generator matrix H of the dual code $C^\perp$ is called a *parity check matrix* of the linear code C. Let G denote a generator matrix of C, then it follows from the definition that

$$GH^T = 0_{k \times (n-k)},$$

where $0_{k \times (n-k)}$ is the $k \times (n-k)$ zero matrix, H^T is the *transpose* of H. Thus, a vector **u** of F^n is a codeword of C if and only if $\mathbf{u}H^T = \mathbf{0}$. For an $[n, k]$ linear code the *rate* R is defined to be k/n. This is a measure of the efficiency of the code.

The Hamming weight of a vector **u** is the number of its nonzero components, and is denoted by wt(**u**). For example, wt(10010) $= 2$. It is easy to prove that the minimum distance of a linear code is equal to the minimum nonzero Hamming weight of the code.

The minimum distance of a code reflects the random-error correcting capacity of the code. Any code with minimum distance $d \geq 2t + 1$ can theoretically correct t random errors. Under the assumption that for each coded message block there are at most t transmission errors, a brute force decoding procedure is to compute the Hamming distance between a received vector and the codewords in turn. If a codeword having a minimum distance from the received vector less than or equal to t is detected, then the original encoded message must be this codeword. By assumption there is one and only one codeword satisfying this property, so the decoding procedure must terminate. However, this decoding procedure, which applies theoretically to every block code, is computationally infeasible when the number of codewords is large. Thus, it is important to construct (n, M, d) codes over a finite field $GF(q)$ having the following properties:

1. the rate $R = (\log_q M)/n$ is large;

2. the minimum distance is large;

3. there are efficient encoding and decoding algorithms.

There is a trade-off between the rate R and the minimum distance d for any linear code. If C is an $[n, k, d]$ linear code, then $r = n - k$ is the rank of H. Note when speaking of $[n, k, d]$ linear codes, we indicate also the minimum distance. Thus, we have the following Singleton bound

$$n - k \geq d - 1.$$

By definition and the Singleton bound

$$R + \frac{d}{n} \leq 1 + \frac{1}{n}.$$

This means that we have to make a compromise between the rate and the minimum distance. The third requirement above is important, as codes with a good rate and good minimum distance may be practically useless if there are no efficient encoding and decoding algorithms.

From the rate and minimum distance viewpoint, codes satisfying $n - k = d - 1$ are attractive. Such codes are called *maximum distance separable* (briefly, MDS).

For linear codes a general but still inefficient decoding method is the maximum likelihood decoding. Let C be an $[n, k]$ linear code over a finite field $GF(q)$. For any vector $\mathbf{u}$ the set

$$\mathbf{u} + C = \{\mathbf{u} + x | x \in C\}$$

is called a *coset* of C. Every vector of $GF(q)^n$ is contained in such a coset. Since C is a linear subspace of $GF(q)^n$, two cosets are either disjoint or identical.

Suppose that an encoded block $\mathbf{x}$ was sent via a noisy channel, and a vector $\mathbf{y}$ is received, then the error vector is $\mathbf{e} = \mathbf{y} - \mathbf{x}$. Let H denote the parity check matrix of the linear code. It follows that

$$\mathbf{e}H^T = \mathbf{y}H^T,$$

which is called the *syndrome* of $\mathbf{y}$. The syndrome is a row vector of length $n - k$. Obviously, all the vectors of a coset have the same syndrome. The minimum weight vector of a coset is called the *coset leader* (if there is more than one vector with the same minimum weight, choose one at random and take it as the coset leader). With the maximum likelihood decoding we first compute the syndrome of a received vector $\mathbf{y}H^T$, then take the coset leader $\mathbf{e}$ of the coset having syndrome $\mathbf{y}H^T$ as the error vector. Finally the received block is decoded to be $\mathbf{y} - \mathbf{e}$. This is of course not a complete decoding procedure since decoding errors could occur. In addition, this maximum likelihood decoding is quite inefficient when the parameter $n - k$ is large since the number of possible syndromes is q^{n-k} for an $[n, k]$ linear code over $GF(q)$.

An important class of linear codes is *cyclic codes*. A code C is cyclic if it is linear and any cyclic shift of a codeword is also a codeword, i.e., $(c_{n-1}, c_0, ..., c_{n-2})$ is a codeword whenever $(c_0, c_1, ..., c_{n-1})$ is in C. $C = \{000, 110, 101, 011\}$ is a simple cyclic code.

To study cyclic codes, it is convenient to set up a one-to-one correspondence between the set of cyclic codes of length n over a field F and the set of ideals of the residue class ring $F[x]_n = F[x]/(x^n - 1)$, which consists of the residue classes modulo $x^n - 1$. Since polynomials of degree less than or equal to $n - 1$ belong to different residue classes, we use these polynomials to represent their residue classes. Thus, the ring $F[x]_n$ consists of all polynomials of $F[x]$ with degree no more than $n - 1$, where the multiplication of two polynomials is carried out modulo $x^n - 1$.

Let $c(x) = c_0 + c_1 x + \cdots + c_{n-1}x^{n-1}$ be an element of $F[x]_n$. Then the product of x and $c(x)$ in $F[x]_n$ is

$$
\begin{aligned}
xc(x) &= c_0 x + c_1 x^2 + \cdots + c_{n-1}x^n \\
&= c_{n-1} + c_0 x + \cdots + c_{n-2}x^{n-2}.
\end{aligned}
$$

Recall that an *ideal* I of a commutative ring R is a subring of R such that $ri \in I$ for any $r \in R$ and any $i \in I$. Define the natural one-to-one correspondence between F^n and $F[x]_n$ by

$$
\tau : \; \mathbf{c} = (c_0, c_1, ..., c_{n-1}) \mapsto c(x) = c_0 + c_1 x + \cdots + c_{n-1}x^{n-1}.
$$

Then it is easily seen that $C \subseteq F^n$ is a cyclic code if and only if $\tau(C)$ is an ideal of $F[x]_n$. Thus, a cyclic code of length n over F can be defined as an ideal of $F[x]_n$. We shall often identify a cyclic code C and its image $\tau(C)$, an ideal of $F[x]_n$.

By definition a principal ideal I of a commutative ring R is generated by a single element, i.e., $I = (a)$. It is well known that $F[x]_n$ is a *principal ideal domain* (briefly, PID), i.e. every ideal of $F[x]_n$ is principal. This means that every cyclic code C of length n consists of all multiples of a fixed polynomial $g(x)$, which is called a *generator polynomial*. If C is a nonzero ideal of $F[x]_n$, i.e., a cyclic code of length n, let $g(x)$ be the unique monic polynomial $g(x)$ of minimum degree in C. Let $r = \deg(g(x))$ and suppose that $c(x) \in C$. Write $c(x) = q(x)g(x) + r(x)$, where $\deg(r(x)) < r$. By the linearity of the code, $r(x) = c(x) - q(x)g(x) \in C$. It follows that $r(x) = 0$. Thus, $g(x)$ is a generator polynomial of the code C, and every $c(x) \in C$ can be written uniquely as $c(x) = g(x)f(x)$, where $f(x)$ is a polynomial of $F[x]$ with degree less than $n - r$. In addition, the dimension of C is $n - r$.

If $g(x) = g_0 + g_1 x + \cdots + g_r x^r$, then the code C, when considered as a subspace of F^n, has the $(n - r) \times n$ generator matrix

$$G = \begin{bmatrix} g_0 & g_1 & g_2 & \cdots & g_r & 0 & 0 & \cdots & 0 \\ 0 & g_0 & g_1 & g_2 & \cdots & g_r & 0 & \cdots & 0 \\ 0 & 0 & g_0 & g_1 & g_2 & \cdots & g_r & \cdots & 0 \\ \vdots & \vdots & \vdots & \vdots & \vdots & \vdots & \vdots & \vdots & \vdots \\ 0 & 0 & 0 & 0 & g_0 & g_1 & g_2 & \cdots & g_r \end{bmatrix}.$$

Let C be a cyclic code of length n with the generator polynomial $g(x)$. Note that $g(x)$ divides $x^n - 1$. Then $h(x) = (x^n - 1)/g(x)$ is called the *check polynomial* of C, since $c(x) \in F[x]_n$ is a codeword of C if and only if $c(x)h(x) = 0$, where the multiplication of the two polynomials is carried out in $F[x]_n$, i.e., the usual multiplication modulo $x^n - 1$.

Let $h(x) = \sum_{i=0}^{n-r} h_i x^i$. The parity check matrix of the code C is described by the following $r \times n$ matrix

$$H = \begin{bmatrix} 0 & 0 & 0 & 0 & h_{n-r} & \cdots & h_2 & h_1 & h_0 \\ 0 & 0 & 0 & h_{n-r} & \cdots & & h_2 & h_1 & h_0 & 0 \\ 0 & 0 & h_{n-r} & \cdots & & h_2 & h_1 & h_0 & 0 & 0 \\ \vdots & \vdots & \vdots & \vdots & \vdots & \vdots & \vdots & \vdots & \vdots \\ h_{n-r} & \cdots & h_2 & h_1 & h_0 & 0 & 0 & \cdots & 0 \end{bmatrix}.$$

Since a generator polynomial of a cyclic code of length n over F always divides $x^n - 1$, it is necessary to consider the factorization of the *cyclotomic polynomial* $x^n - 1$ over F. From now on we consider only cyclic codes over finite fields $GF(q)$ in this section. To study the factors of the polynomial $x^n - 1$, we assume that $\gcd(n, q) = 1$. By this assumption the *multiplicative order* of q modulo n, which is the smallest integer m such that n divides $q^m - 1$, exists. It follows that $x^n - 1$ divides $x^{q^m - 1} - 1$, but does not divide $x^{q^s - 1} - 1$ for $0 < s < m$. Thus the zeros of $x^n - 1$, which are called the nth roots of unity, lie in this extension field $GF(q^m)$, but in no smaller extension field. Since

$\gcd(nx^{n-1}, x^n - 1) = 1$ due to $\gcd(n, q) = 1$, the cyclotomic polynomial $x^n - 1$ has n distinct zeros, say $\alpha_0, \alpha_1, ..., \alpha_{n-1}$, and factors over $GF(q^m)$ into

$$x^n - 1 = \prod_{i=0}^{n-1} (x - \alpha_i).$$

It is easily seen that there are nth primitive roots α such that the set $\{\alpha^i | i = 0, 1, ..., n - 1\}$ consists of exactly the n nth roots of unity.

The *cyclotomic coset* containing s is defined by

$$C_s = \{s, sq, sq^2, ..., sq^{m_s-1}\},$$

where m_s is the smallest integer such that $sq^{m_s} = s \bmod n$, and the multiplications are carried out modulo n. Then one can prove that the minimum polynomial of α^s over $GF(q)$, which is defined to be a polynomial of $GF(q)[x]$ having α^i as a root and minimum degree, is given by

$$M^{(s)}(x) = \prod_{i \in C_s} (x - \alpha^i).$$

Since the cyclotomic cosets form a partition of the set $\{0, 1, ..., n - 1\}$, we have

$$x^n - 1 = \prod_s M^{(s)}(x),$$

where s ranges over a set of coset representatives modulo n. This is the factorization of $x^n - 1$ into irreducible polynomials over $GF(q)$.

One important class of cyclic codes is the class of *BCH codes*. A cyclic code of length n over $GF(q)$ is a BCH code of designed distance $\delta \geq 2$ if, for some integer $b \geq 0$, its generator polynomial

$$g(x) = \operatorname{lcm}\{M^{(b)}(x), M^{(b+1)}(x), ..., M^{(b+\delta-2)}(x)\}.$$

It can be proven [51, p.203] that the actual minimum distance of the BCH code is at least δ. In such a BCH code c is a codeword if and only if

$$c(\alpha^b) = c(\alpha^{b+1}) = \cdots = c(\alpha^{b+\delta-2}) = 0.$$

Consider now a specific BCH code. First construct the finite field $GF(2^4)$ defined by $\alpha^4 + \alpha + 1 = 0$. In this field the minimum polynomials of $\alpha, \alpha^2, \alpha^4$ and α^8 are

$$M^{(1)} = M^{(2)} = M^{(4)} = M^{(8)} = x^4 + x + 1;$$

and those of $\alpha^3, \alpha^6, \alpha^{12}$ and α^9 are

$$M^{(3)} = M^{(6)} = M^{(12)} = M^{(9)} = x^4 + x^3 + x^2 + x + 1.$$

Let

$$
\begin{aligned}
g(x) &= \text{lcm}\{\alpha, \alpha^2, \alpha^3, \alpha^4\} \\
&= (x^4 + x + 1)(x^4 + x^3 + x^2 + x + 1) \\
&= x^8 + x^7 + x^6 + x^4 + 1.
\end{aligned}
$$

The corresponding binary [15, 7] BCH code generated by $g(x)$ has the designed minimum distance $\delta = 5$. Since $g(x)$ is itself a codeword having Hamming weight 5, the minimum distance of this code is 5. Thus, it is a double-error-correcting BCH code.

So far we have introduced some notions and notations as well as some basic codes which we need for the investigation of codes based on the Chinese Remainder Theorem. For further details about block codes we refer to [51].

6.2   Redundant Residue Codes

Recall first the modular computing techniques in Chapter 3. To make those modular algorithms work properly, the moduli m_i should be chosen such that the norm of $m = \prod_i m_i$ is larger than that of the arithmetic expression to be computed, where the norm for $\mathbf{Z}$ is the absolute value for integers, and the norm for $\mathbf{Z}[x]$ is the degree for polynomials. This is to ensure that the images of the arithmetic expression can provide enough information about the value of the original arithmetic expression. But requiring that the norm of m is much larger than that of the arithmetic expression to be computed is not only useless, but also harmful to the efficiency of a modular algorithm for the computation of the value of the arithmetic expression. However, for secret sharing (see Section 7.1) and error-correcting purpose, this kind of redundancy in the moduli is necessary.

To show the idea of applying the CRT in constructing error-correcting codes, we take a small example. The four polynomials

$$
\begin{aligned}
m_0(x) &= 1 + x + x^3, \quad m_1(x) = 1 + x^2 + x^3, \\
m_2(x) &= 1 + x^3, \quad m_3(x) = x^3
\end{aligned}
$$

are pairwise relatively prime over $GF(2)$. For each polynomial $c(x) = c_0 + c_1 x + \cdots + c_5 x^5 \in GF(2)[x]_6$, let

$$r_i(x) = c(x) \bmod m_i(x), \quad \text{for } i = 0, 1, 2, 3.$$

By the CRT any two of the four residues $r_0(x), r_1(x), r_2(x)$ and $r_3(x)$ are sufficient to recover the original polynomial $c(x)$. Thus, each message can be divided
into blocks of length 6 and each message block $c(x)$ can be encoded as a codeword $c'(x) = r_0(x) + r_1(x)x^3$. By the CRT the set

$$\{c'(x) = r_0(x) + r_1(x)x^3 | r_i(x) = c(x) \bmod m_i(x), c(x) \in GF(2)[x]_6\}$$

forms a $[6,6,1]$ linear code over $GF(2)$ without any random-error correcting ability. For $c(x) = x + x^2 + x^4$, the corresponding two residues are
$r_0(x) = 0, r_1(x) = 1$. It follows that x^3 is a codeword, and therefore the
minimum distance of the $[6,6]$ linear code is one. However, if we add redundancy by considering the redundant residues $r_2(x) = c(x) \bmod m_2(x)$ and
$r_3(x) = c(x) \bmod m_3(x)$ we get the following $[12,6,d]$ linear code

$$\mathcal{C} = \{r_0(x) + r_1(x)x^3 + r_2(x)x^6 + r_3(x)x^9 | c(x) \in GF(2)[x]_6\},$$

where $4 \geq d \geq 3$ and $r_i(x) = c(x) \bmod m_i(x)$. This linear code $\mathcal{C}$ can correct
one random error. The minimum distance is no less than three since the original
polynomial $c(x)$ must be zero if any two of the four residues are zero polynomials.
On the other hand, for $c(x) = x + x^2 + x^4$ the corresponding $c'(x) = x^3 + x^8 +
x^{10} + x^{11}$. Thus, $d \leq 4$.

Though the above linear code $\mathcal{C}$ has no random-error correcting capacity,
it does provide much information for burst errors. A binary *burst error* is of
such a form $(0...01....10...0)$, i.e., all errors occurred consecutively for a number
of places. For instance, let $r(x) = r_0 + r_1x + \cdots + r_{11}x^{11} \in GF(2)[x]$ be a
received word, and assume that only a burst error of length no more than 4
occurred during transmission. Note that there are $13 - i$ possible burst errors
of length i, the total number of possible burst errors under this assumption is
$\sum_{i=1}^{4}(13 - i) = 42$. Since the length of the burst error is at most 4, at least
two of the 4 received residues must be the same as the originally sent ones,
and by the CRT the two residues are sufficient to recover the original message
$c(x)$. Thus, with this linear code many of the 42 possible burst errors can be
corrected.

It is possible to construct linear codes based on the CRT having a reasonable
random-error correcting ability and ideal burst-error correcting capacity. We
shall see that the important Reed-Solomon codes belong to this class.

Redundant residue codes were studied by Stone [90], Bossen and Yau [14],
and Mandelbaum [52]. Those codes can be generally described as follows.

Let $m_0(x), m_1(x), \cdots, m_{s+t-1}(x) \in GF(q)[x]$ be pairwise relatively prime,
with degree $e_0, e_1, \cdots, e_{s+t-1}$ respectively, and $k = \sum_{i=0}^{s-1} e_i$. For each polynomial
$p(x)$ of degree no more than $k - 1$, let

$$r_i(x) = p(x) \bmod m_i(x), \quad i = 0, \cdots, s+t-1.$$

Hereafter let $\mathbf{r}_i = (r_{i,0}, r_{i,1}, ..., r_{i,e_i-1})$ if the residue $r_i(x) = r_{i,0} + r_{i,1}x + \cdots + r_{i,e_i-1}x^{e_i-1}$. The redundant residue codes are described by

$$\mathcal{C} = \{(\mathbf{r}_0, \cdots, \mathbf{r}_{s-1}, \mathbf{r}_s, \cdots, \mathbf{r}_{s+t-1}) | p(x) \in GF(q)[x]_k\}, \qquad (6.1)$$

where $GF(q)[x]_k$ is the set of all polynomials of degree no more than $k-1$ over $GF(q)$. By the CRT the first s residues $r_0(x), r_1(x), \cdots, r_{s-1}(x)$ are sufficient to recover the original polynomial $p(x)$. So the last t residues $r_s(x), \cdots, r_{s+t-1}(x)$ are redundant and are used for error correction. It follows from the definition and the CRT that the following conclusion about the redundant residue codes holds.

Theorem 6.2.1 *Let the symbols be the same as before. The redundant residue code described by (6.1) is an $[\sum_{i=0}^{s+t-1} e_i, k]$ linear code, where $k = \sum_{i=0}^{s-1} e_i$.*

In constructing the $[6, 6, 1]$ and $[12, 6, d]$ codes before, the moduli were chosen to have the same degree 3. Generally, the degree of the moduli can be different. To illustrate this, we take another example.

Consider the five polynomials $m_0(x) = x, m_1(x) = x + 1, m_2(x) = x + 2$, $m_3(x) = x^2 + 1, m_4(x) = x^2 + x + 2$ over $GF(3)$. They are pairwise relatively prime. Basing on these moduli, we have the following $[7, 3, 3]$ linear code over $GF(3)$:

$$\mathcal{C} = \left\{ \begin{array}{lllll} 0000000, & 1111010, & 2222020, & 0210101, & 0120202, \\ 0111021, & 0222012, & 1021111, & 2012222, & 1201212, \\ 2102121, & 1220011, & 2110022, & 1002001, & 2001002, \\ 0022110, & 0011220, & 0201122, & 0102211, & 1100120, \\ 2200210, & 1122200, & 2211100, & 2020112, & 1010221, \\ 1212102, & 2121201 \end{array} \right\}.$$

In this code there are 2 codewords having Hamming weight 3, 6 codewords having Hamming weight 4, 12 codewords having Hamming weight 5, and 6 codewords having Hamming weight 6.

Generally speaking, the minimum distance of the redundant residue code $\mathcal{C}$ described by (6.1) is not easy to determine. It can be done in certain special cases, as shown in later sections.

6.3 Reed-Solomon Codes

A very important class of redundant residue codes is the Reed-Solomon codes. They are both theoretically and practically important. The minimum distance and dimension of the class of redundant residue codes are easy to derive.

Let α be a generating element of the finite field $GF(q)$, and $n = q - 1$. The n polynomials $m_i(x) = x - \alpha^i$ are pairwise relatively prime, where $0 \leq i \leq n-1$. Let $r_i(x) = u(x) \bmod m_i(x)$ for $i = 0, \cdots, n - 1$. The Reed-Solomon code is defined by

$$C = \{(r_0, r_1, \cdots, r_{n-1}) | u(x) \in GF(q)[x]_k\}, \tag{6.2}$$

where r_i and $GF(q)[x]_k$ are defined as before. Since the moduli here are of degree one, $r_i = u(\alpha^i)$ for $0 \leq i \leq n - 1$. Thus, the code can also be described as

$$C = \{(u(1), u(\alpha), \cdots, u(\alpha^{n-1})) | u(x) \in GF(q)[x]_k\}. \tag{6.3}$$

On the one hand, a polynomial of degree no larger than $k - 1$ over $GF(q)$ has at most $k - 1$ zeros, thus the Hamming weight of each codeword is no less than $n - k + 1$. On the other hand, the polynomial $u(x) = (x - 1)(x - \alpha) \cdots (x - \alpha^{k-2})$ over $GF(q)$ has exactly $k - 1$ zeros over $GF(q)$. Thus, the minimum distance of the Reed-Solomon code of (6.2) has the minimum distance $n - k + 1$. An $[n, k, d]$ code over $GF(q)$ is called *maximum-distance separable (MDS)* if $d = n - k + 1$. By definition, the dimension of the Reed-Solomon code is k. In summary, we have the following conclusion.

Theorem 6.3.1 *The Reed-Solomon code defined by (6.2) is an $[n, k, n - k + 1]$ linear MDS code.*

Similar to other redundant residue codes, the Reed-Solomon code of (6.2) can be encoded as follows. Let $\mathbf{u} = (u_0, \cdots, u_{k-1})$, be the message symbols to be encoded, and set

$$u(x) = \sum_{i=0}^{k-1} u_i x^i.$$

Then the codeword corresponding to $\mathbf{u}$ is

$$\mathbf{c} = (u(1), u(\alpha), \cdots, u(\alpha^{n-1})). \tag{6.4}$$

Cyclic codes are relatively easy to encode and decode. Redundant residue codes are usually not cyclic, but Reed-Solomon codes are cyclic and belong to the BCH class.

Theorem 6.3.2 *The Reed-Solomon code defined by (6.2) or (6.3) is an $[n, k, n - k + 1]$ BCH code with generator polynomial*

$$g(x) = (x - \alpha)(x - \alpha^2) \cdots (x - \alpha^{d-1}),$$

where $d = n - k + 1$.

Proof: The first step of our proof is to show that for each codeword

$$\mathbf{c} = (u(1), u(\alpha), ..., u(\alpha^{n-1})),$$

where $u(x) \in GF(q)[x]_k$, the polynomial $c(x) = \sum_{i=0}^{n-1} c_i x^i$ has zeros $\alpha, \alpha^2, ...,$
α^{d-1}. Note that

$$
\begin{aligned}
c(\alpha^r) &= \sum_{i=0}^{n-1} u(\alpha^i) \alpha^{ir} \\
&= \sum_{i=0}^{n-1} \left(\sum_{j=0}^{k-1} u_j \alpha^{ij} \right) \alpha^{ir} \\
&= \sum_{j=0}^{k-1} u_j \left(\sum_{i=0}^{n-1} \alpha^{i(j+r)} \right)
\end{aligned}
\tag{6.5}
$$

and

$$\sum_{i=0}^{n-1} \beta^i = 0 \tag{6.6}$$

holds for each nth root of unity $\beta \neq 1$. If $1 \leq r \leq d-1$ and $0 \leq j \leq k-1$,
then $1 \leq j+r \leq d-1+k-1 = n-1$. Thus $\alpha^{j+r} \neq 1$, is an nth root of unity.
By (6.5) and (6.6) $c(\alpha^r) = 0$ for each r with $1 \leq r \leq d-1$. Hence for each
codeword $\mathbf{c}$ the polynomial $g(x)$ divides the corresponding polynomial $c(x)$.

The second step is to prove that each of $1, \alpha^d, \alpha^{d+1}, ..., \alpha^{n-1}$ is not a zero
of some codeword. Consider first the polynomial $u(x) = u_0 \neq 0$. For the
corresponding codeword $\mathbf{c}$ we have

$$c(1) = n u_0 \neq 0.$$

For any α^r with $d \leq r \leq n-1$ we have $1 \leq n-r \leq k-1$. Consider now the
polynomial $u(x) = x^j$, where $j = n-r$. For the corresponding codeword $\mathbf{c}$ we
have $c(\alpha^r) = n \neq 0$. This completes the second step of the proof.

By the proof of the above two steps, $g(x)$ is the generator polynomial of the
Reed-Solomon codes. The remaining conclusions are straightforward. $\qquad\square$

The following theorem summarizes some interesting properties of the dual
codes of the Reed-Solomon codes.

Theorem 6.3.3 *The dual code of the $[n, k, n-k+1]$ Reed-Solomon code of
(6.2) is an $[n, n-k, k+1]$ BCH MDS code with generator polynomial*

$$g'(x) = (x-1)(x-\alpha)\cdots(x-\alpha^{k-1}),$$

but not a Reed-Solomon code of form (6.2).

Proof: Let $\mathcal{C}^{\perp}$ denote the dual code of the $[n, k, n-k+1]$ Reed-Solomon
code of (6.2). By definition $\mathbf{v} = (v_0, v_1, ..., v_{n-1}) \in \mathcal{C}^{\perp}$ iff for any polynomial
$u(x) \in GF(q)[x]_k$

$$\sum_{i=0}^{n-1} v_i u(\alpha^i) = \sum_{j=0}^{k-1} u_j \sum_{i=0}^{n-1} v_i \alpha^{ij} = 0,$$

which is equivalent to

$$v(\alpha^j) = \sum_{i=0}^{n-1} v_i \alpha^{ij} = 0, \ j = 0, 1, ..., k-1.$$

Thus for every codeword $\mathbf{v} \in \mathcal{C}^\perp$ the corresponding polynomial $v(x) = v_0 + v_1 x + \cdots + v_{n-1} x^{n-1}$ is divisible by $g'(x)$. Since the dimension of $\mathcal{C}^\perp$ is $n-k$, $g'(x)$ must be the generator polynomial of $\mathcal{C}^\perp$.

Note that the generator polynomial $g'(x)$ has k consecutive powers α^0, α^1, ..., α^{k+1-2} as roots. It follows from the BCH bound mentioned before that the minimum distance d' of $\mathcal{C}^\perp$ satisfies

$$d' \geq k+1.$$

On the other hand, by the Singleton bound

$$d' \leq n - (n-k) + 1 = k+1.$$

Combining the above two inequalities gives $d' = k+1$. Thus, $\mathcal{C}^\perp$ is an $[n, n-k, k+1]$ BCH MDS code. Finally, by Theorem 6.3.2 $\mathcal{C}^\perp$ is not a Reed-Solomon code of form (6.2). $\qquad\qquad\square$

One way to construct codes is to extend a code $\mathcal{C}$ by adding an overall parity check

$$c_n = -\sum_{i=0}^{n-1} c_i$$

to each codeword $\mathbf{c} = (c_0, c_1, ..., c_{n-1})$ of $\mathcal{C}$. With this method an $[n, k]$ linear code is extended into an $[n+1, k]$ linear code. The extended code of the $[4, 2, 1]$ linear code over $GF(2)$ $\mathcal{C} = \{0000, 1110, 0001, 1111\}$ is the $[5, 2, 2]$ linear code

$$\{00000, 11101, 00011, 11110\}.$$

In this example the extended code has the minimum distance greater than that of the original one. However, adding an overall parity check to a code does not always increase the minimum distance. But this is true for Reed-Solomon codes.

Theorem 6.3.4 *Let $\mathcal{C}$ be the $[n, k, n-k+1]$ Reed-Solomon code defined by (6.2). Then extending each codeword $\mathbf{c} = (c_0, c_1, ..., c_{n-1})$ of $\mathcal{C}$ by adding an overall parity check $c_n = -\sum_{i=0}^{n-1} c_i$ produces an $[n+1, k, n-k+2]$ MDS code, denoted by $\mathcal{E}$.*

Proof: Let $d = n - k + 1$ and

$$\mathbf{c} = (u(1), u(\alpha), ..., u(\alpha^{n-1})) \in \mathcal{C},$$

where $u(x) = u_0 + u_1 x + \cdots + u_{k-1} x^{k-1} \in GF(q)[x]_k$. We first prove that $u_0 \neq 0$ if $\mathbf{c} \in \mathcal{C}$ has Hamming weight d. Assume $\text{wt}(\mathbf{c}) = d$ and $\alpha^{i_1}, \alpha^{i_2}, ..., \alpha^{i_{n-d}}$ are the distinct roots of $u(x)$ in $GF(q)$.

Suppose that $u_0 = 0$, then there is an $u_1(x) \in GF(q)[x]$ such that $u(x) = x u_1(x)$. Apparently, $u_1(x)$ has also zeros $\alpha^{i_1}, \alpha^{i_2}, ..., \alpha^{i_{n-d}}$. Choose an $i \in \{0, 1, ..., n-1\} \setminus \{i_1, i_2, ..., i_{n-d}\}$ and consider the polynomial

$$v(x) = (x - \alpha^i) u_1(x).$$

Clearly, $\deg(v(x)) = \deg(u(x)) \leq k-1$. Hence $\mathbf{h} = (v(1), v(\alpha), ..., v(\alpha^{n-1})) \in \mathcal{C}$. But $\text{wt}(\mathbf{h}) = \text{wt}(\mathbf{c}) - 1 = d - 1$, which is contrary to the minimality of the minimum distance d. Thus, $u_0 \neq 0$.

For the codeword $\mathbf{c} = (u(1), u(\alpha), ..., u(\alpha^{n-1})) \in \mathcal{C}$, the parity check

$$
\begin{aligned}
c_n &= -\sum_{i=0}^{n-1} u(\alpha^i) \\
&= -\sum_{j=0}^{k-1} u_j \sum_{i=0}^{n-1} \alpha^{ij} \\
&= -n u_0.
\end{aligned}
$$

It follows from the above proved conclusion that $c_n \neq 0$ if $\text{wt}(\mathbf{c}) = d$. Thus the minimum distance of the extended Reed-Solomon code $\mathcal{E}$ is $d + 1 = n - k + 2$. The remaining conclusions are straightforward.   □

When defining the Reed-Solomon codes—a special class of redundant residue codes based on the CRT—we employed the moduli $m_i(x) = (x - \alpha^i)$ with $0 \leq i \leq n - 1$, where $n = q - 1$. There is one more modulus $m_n(x) = x$ of degree one which can be used to extend the Reed-Solomon code of (6.2). This extended code is described by

$$\mathcal{L} = \{(\mathbf{r}_0, \mathbf{r}_1, ..., \mathbf{r}_{n-1}, \mathbf{r}_n) | u(x) \in GF(q)[x]_k\}, \tag{6.7}$$

where $r_i(x) = u(x) \bmod m_i(x)$ for $i = 0, 1, ..., n$. Since all the $n + 1$ moduli are of degree one, this code is also described by

$$\mathcal{L} = \{(u(1), u(\alpha), ..., u(\alpha^{n-1}), u(0)) | u(x) \in GF(q)[x]_k\}. \tag{6.8}$$

It follows from $n + 1 = q$ that $-n u_0 = u_0$ for each $u_0 \in GF(q)$. Thus, $\mathcal{E} = \mathcal{L}$. This proves the following theorem.

Theorem 6.3.5 *The extended Reed-Solomon codes are also redundant residue codes, as described by (6.7).*

There are efficient algorithms for decoding BCH codes (see [51, Section 6, Chapter 9] for example). Those efficient algorithms can be used to decode the Reed-Solomon codes since they are special BCH codes. There is a majority logic decoding method for the codes, which theoretically applies to a number of redundant residue codes. The original majority logic decoding method of Reed and Solomon is of theoretical interest, even though it is usually impractical.

The idea of the majority logic decoding method comes simply from the construction of the Reed-Solomon codes. Suppose that the codeword of (6.4) has been sent, an error vector $\mathbf{e} = (e_0, e_1, ..., e_{n-1})$ has occurred, and $\mathbf{y} = (y_0, y_1, ..., y_{n-1})$ has been received. If there are no errors, i.e., $\mathbf{e} = \mathbf{0}$, any k elements of $\{y_0, y_1, ..., y_{n-1}\}$ can be used to recover the message $\mathbf{u} = (u_0, ..., u_{k-1})$ with the efficient CRAs. Thus there are $\binom{n}{k}$ choices, or *votes*, for the correct $\mathbf{u}$.

If there are errors, some sets of k elements of $\{y_0, y_1, ..., y_{n-1}\}$ give the wrong $\mathbf{u}$. It is not difficult to see that an incorrect $\mathbf{u}$ receives at most $\binom{w+k-1}{k}$ votes, and the correct $\mathbf{u}$ receives at least $\binom{n-w}{k}$ votes, if w errors have occurred.

Thus the message $\mathbf{u}$ will be obtained correctly if

$$\binom{n-w}{k} > \binom{w+k-1}{k},$$

i.e., if $n - w > w + k - 1$ or $d = n - k + 1 > 2w$. So error vectors of weight less than half the minimum distance can be corrected.

Under the assumption that less than $d/2$ errors occur within one received word, the main step of the majority logic decoding method is to compute the $\binom{n}{k}$ solutions of $\mathbf{u}$ with the CRA or with some other method. The second step is to count the votes. The one receiving the largest number of votes must be the original message. Apparently, this method is impractical if $\binom{n}{k}$ is large.

Binary burst errors were discussed in Section 6.3. Generally, a *burst* of length l is a vector whose only nonzeros are among l successive components, the first and the last of which are nonzero. A Reed-Solomon code over $GF(q^m)$ can be used to construct codes over $GF(q)$ which have usually a good burst-error correcting ability.

When the finite field $GF(q)$, where $q = p^m$, is regarded as a vector space over $GF(p)$, it has dimension m. Let $\alpha_0, ..., \alpha_{m-1}$ be a basis of $GF(q)$ over $GF(p)$. Then each element x of $GF(q)$ can be represented as

$$x = \sum_{i=0}^{m-1} x_i \alpha_i,$$

where $x_i \in GF(p)$ for each i. Thus with any basis each element of $GF(q)$ can be represented by an m-tuple of elements from $GF(p)$. If we replace each component of each codeword of the Reed-Solomon code over $GF(q)$ by an m-tuple of elements from $GF(p)$, we obtain an $[n' = mn, k' = mk, d' \geq d]$ linear code over $GF(p)$. This mapping sends linear codes to linear codes, but cyclic codes need not go into cyclic codes. The extended Reed-Solomon codes can be similarly mapped into subfield codes.

Binary codes obtained from Reed-Solomon codes are particularly suited to correcting several bursts. For a binary burst of length l can affect at most r adjacent symbols from $GF(2^m)$, where r is given by

$$(r - 2)m + 2 \leq l \leq (r - 1)m + 1.$$

So if d is much greater than r, many bursts can be corrected.

Reed-Solomon codes can also be used to construct the so-called Justesen codes [45]. It has also been shown that in an $[n, k, d = n - k + 1]$ MDS code over $GF(q)$ the number of codewords of weight w is

$$A_w = \binom{n}{w} \sum_{j=0}^{w-d} (-1)^j \binom{w}{j} (q^{w-d+1-j} - 1).$$

For a proof of this result we refer to [51, pp.320-321]. If Reed-Solomon codes are used for burst-error correction, the algorithm of Section 6.5 also applies.

6.4   Redundant Residue Codes of Degree 2

In this section we shall construct a class of redundant residue codes and derive their parameters. To this end, we need not only polynomials of degree one, but also quadratic irreducible polynomials as our moduli. Thus, the number of monic irreducible polynomials of degree two is essential to our investigation.

The number of monic irreducible polynomials of degree m over $GF(q)$ can be expressed in terms of the Möbius function, which is defined by

$$\mu(n) = \begin{cases} 1 & \text{if } n = 1, \\ (-1)^r & \text{if } n \text{ is the product of } r \text{ distinct primes}, \\ 0 & \text{otherwise}. \end{cases}$$

The following properties of the Möbius function are fundamental:

1. $\sum_{d|n} \mu(d) = \begin{cases} 1 & \text{if } n = 1, \\ 0 & \text{if } n > 1. \end{cases}$

2. If $f(n) = \sum_{d|n} g(d)$ for all positive integers n, then

$$g(n) = \sum_{d|n} \mu(d) f(n/d).$$

This is the Möbius inverse formula.

3. $\Phi(n) = \sum_{d|n} d\mu(n/d)$ for $n \geq 1$, where Φ is the Euler function.

With the Möbius function the number of monic irreducible polynomials of degree m over $GF(q)$, denoted by $I_q(m)$, is described by the following lemma whose proof can be found in [8] or [51, p.115].

Lemma 6.4.1

$$I_q(m) = \frac{1}{m} \sum_{d|m} \mu(d) q^{m/d}.$$

It follows from Lemma 6.4.1 that the number of monic irreducible polynomials of degree two is

$$I_q(2) = \frac{q(q-1)}{2}.$$

Recall the $q-1$ moduli $m_i = x - \alpha^i$ with $0 \leq i \leq q - 2$ for constructing the Reed-Solomon codes over $GF(q)$. In addition to these moduli, we need a number of irreducible quadratics $x^2 + ax + b$ over $GF(q)$ to construct new redundant residue codes. Let $h_1(x), ..., h_t(x)$ be any t monic irreducible polynomials of degree two over $GF(q)$, where $1 \leq t \leq q(q-1)/2$. Then the polynomials $m_i(x)$ for $i = 0, ..., q - 2$ and $h_1(x), ..., h_t(x)$ are pairwise relatively prime. We now define a class of redundant residue codes as

$$Q = \{(\mathbf{r_0}, ..., \mathbf{r_{q-2}}; \mathbf{s_1}, ..., \mathbf{s_t}) | u(x) \in GF(q)[x]_k\}, \qquad (6.9)$$

where $r_i(x) = u(x) \bmod m_i$ for $i = 0, ..., q - 2$, $s_i(x) = u(x) \bmod h_i(x)$, $\mathbf{s_i}$ is a vector of two components corresponding to the quadratic polynomial $s_i(x)$, where $1 \leq i \leq t$.

Theorem 6.4.2 *The code of (6.9) is an $[q - 1 + 2t, k, d \geq q - k + t]$ linear code over $GF(q)$.*

Proof: By definition and the CRT it is an $[q - 1 + 2t, k]$ linear code over $GF(q)$. What remains to be proved is the lower bound for the minimum distance. Consider the nonzero vector

$$\mathbf{c} = (\mathbf{r_0}, ..., \mathbf{r_{q-2}}; \mathbf{s_1}, ..., \mathbf{s_t}) \in Q.$$

Assume that the Hamming weight of the vector $(r_0, ..., r_{q-2})$ is w. Then there are exactly $q - 1 - w$ zero residues among the set $\{r_0, ..., r_{q-2}\}$. It follows that there are at most $\lceil (k - q + 1 + w)/2 \rceil - 1$ zero residues among $s_1(x), ..., s_t(x)$, otherwise by the CRT the original polynomial $u(x)$ must be zero and therefore the codeword c is zero. Thus, the Hamming weight of the vector $(s_1, ..., s_t)$ is at least $t - \lceil (k - q + 1 + w)/2 \rceil + 1$. Hence

$$
\begin{aligned}
\text{wt}(\mathbf{c}) \ &\geq\ w + t - \left\lceil \frac{k - q + 1 + w}{2} \right\rceil + 1 \\
&\geq\ w + t + 1 - \frac{k - q + 2 + w}{2} \\
&=\ t + \frac{w + q - k}{2}.
\end{aligned}
$$

Note that the minimum distance of the Reed-Solomon $[q - 1, k, q - k]$ code is $q - k$, we have $w \geq q - k$ and

$$
\begin{aligned}
\text{wt}(\mathbf{c}) \ &\geq\ t + \frac{w + q - k}{2} \\
&\geq\ q - k + t.
\end{aligned}
$$

Hence

$$
\text{wt}(\mathbf{c}) \geq q - k + t. \tag{6.10}
$$

In addition, by the Singleton bound

$$
d \leq q - k + 2t. \tag{6.11}
$$

Combining (6.10) and (6.11) gives

$$
q - k + t \leq d \leq q - k + 2t. \tag{6.12}
$$

This completes the proof.   □

The following is a corollary of Theorem 6.4.2.

Theorem 6.4.3 *Let $m_i(x)$ be the same as before for $i = 0, 1, ..., q - 2$ and $m(x)$ be any quadratic irreducible polynomial over $GF(q)$. Define*

$$
\mathcal{R} = \{(r_0, ..., r_{q-2}, r) | u(x) \in GF(q)[x]_k\},
$$

where r_i are defined as before, $r(x) = u(x) \bmod m(x)$ and r is the pair corresponding to the polynomial $r(x)$. Then $\mathcal{R}$ is a $[q + 1, k, d]$ linear code, where $d = q - k + 1$ or $q - k + 2$.

Proof: The minimum distance follows from (6.12) and other conclusions from Theorem 6.4.2.   □

This $[q + 1, k, d]$ redundant residue code is almost MDS, since we have $d = q - k + 2$ or $q - k + 1$. If we add another redundant symbol by considering the modulus $m_{q-1}(x) = x$, we have the following code

$$\mathcal{Q}' = \{(\mathbf{r}_0, ..., \mathbf{r}_{q-2}, \mathbf{r}_{q-1}; \mathbf{s}_1, ..., \mathbf{s}_t) | u(x) \in GF(q)[x]_k\}, \tag{6.13}$$

where all symbols are the same as in (6.9) except that r_{q-1} is defined by $r_{q-1}(x) = u(x) \bmod x$. Since the extended Reed-Solomon code is MDS, with arguments similar to the proof of Theorem 6.4.2 one can prove the following theorem.

Theorem 6.4.4 *The code $\mathcal{Q}'$ is a $[q + 2t, k, d \geq q - k + t + 1]$ linear code over $GF(q)$.*

As a corollary of Theorem 6.4.4, we have the following conclusion.

Theorem 6.4.5 *Let $m_i(x)$ be defined as before for $i = 0, 1, ..., q - 1$ and $m(x)$ be any quadratic irreducible polynomial over $GF(q)$. Define*

$$\mathcal{R}' = \{(\mathbf{r}_0, ..., \mathbf{r}_{q-2}, \mathbf{r}_{q-1}, \mathbf{r}) | u(x) \in GF(q)[x]_k\},$$

where $\mathbf{r}_i$ are defined as before, $r(x) = u(x) \bmod m(x)$ and $\mathbf{r}$ is the pair corresponding to the polynomial $r(x)$. Then $\mathcal{R}$ is a $[q + 2, k, d \geq q - k + 2]$ linear code.

Another interesting case of the two classes of redundant residue codes described by (6.9) and (6.13) is $t = (q - 1)q/2$. This gives the block lengths $(q + 1)(q - 1)$ and q^2 respectively.

6.5 Bossen-Yau Codes

A special class of the redundant residue codes is the Bossen-Yau codes [14], which are instantaneously decodable with a modest amount of hardware consisting of field multipliers and adders for the correcting of burst errors. Those codes are very suitable when instantaneous decoding is desired. In this section we shall generalize the original Bossen-Yau codes and their decoding algorithm.

Let $m_1(x), ..., m_{s+t}(x)$ be $s + t$ pairwise relatively prime polynomials of degree m over $GF(p)$. For any $u(x) \in GF(q)[x]_{ms}$, let $r_i(x)$ be the remainder obtained by dividing $u(x)$ by $m_i(x)$ as before. Let $\mathbf{r}_i = (r_{i,0}, r_{i,1}, ..., i_{r,m-1})$ which corresponds to the residue polynomial $r_i(x) = r_{i,0} + r_{i,1}x + \cdots + r_{i,m-1}x^{m-1}$. Then the Bossen-Yau codes are defined by

$$\mathcal{C} = \{(\mathbf{r}_1, \mathbf{r}_2, ..., \mathbf{r}_{s+t})|u(x) \in GF(q)[x]_{ms}\}, \tag{6.14}$$

which is an $[n = (s+t)m, k = ms]$ linear code over $GF(q)$.

Consider now an example. Let $m_1 = x^2$, $m_2 = x^2+1$ and $m_3(x) = x^2+x+1$, which are pairwise relatively prime over $GF(2)$. The code of (6.14) with respect to these moduli and $s = 2$ and $t = 1$ is the following $[6, 4, 2]$ linear code over $GF(2)$:

$$\mathcal{C}_1 = \left\{ \begin{array}{llll} 000000, & 101010, & 010101, & 001011, \\ 000110, & 111111, & 100001, & 101100, \\ 011110, & 010011, & 011000, & 100111, \\ 111001, & 110100, & 001101, & 110010 \end{array} \right\}. \tag{6.15}$$

Assume that $\alpha_1, ..., \alpha_m$ form a basis of $GF(q^m)$ over $GF(q)$. Define the following mapping from $GF(q)^m$ to $GF(q^m)$

$$\varepsilon : \mathbf{a} = (a_1, ..., a_m) \mapsto \sum_{i=1}^{m} a_i \alpha_i.$$

Apparently, $\varepsilon(\mathbf{a} + \mathbf{b}) = \varepsilon(\mathbf{a}) + \varepsilon(\mathbf{b})$. Under this mapping the $[m(s+t), ms]$ linear code over $GF(q)$ is transformed into the following $[s+t, s]$ linear code over $GF(q^m)$

$$\varepsilon(\mathcal{C}) = \{(\varepsilon(\mathbf{r}_1), ..., \varepsilon(\mathbf{r}_{s+t}))|u(x) \in GF(q)[x]_{ms}\}. \tag{6.16}$$

By the CRT the message $u(x)$ must be zero if s or more residues are zeros. It follows that the minimum distance of this code over $GF(q^m)$ is $t + 1$, and this code is an $[s+t, s, t+1]$ MDS code.

It might be possible that the code $\varepsilon(\mathcal{C})$ can be obtained by rearranging the columns of the $[s+t, s, t+1]$ Reed-Solomon code over $GF(q^m)$. If this is true, the Bossen-Yau code described by (6.14) is a column-rearrangement of the subfield code of the $[s+t, s, t+1]$ Reed-Solomon code over $GF(q^m)$.

A decoding procedure developed by Bossen and Yau is based on the Chinese Remainder Theorem and may be generalized as follows. Let $m_1, ..., m_{s+t}$ be pairwise relatively prime polynomials of degree m over $GF(q)$, $u(x)$ be a polynomial of degree less than s over $GF(q)$, and

$$r_i(x) = u(x) \bmod m_i(x), \quad i = 1, ..., s+t.$$

For $i = 1, ..., s$, set $M_i'(x) = (\prod_{j=1}^{s} m_j(x))/m_i(x)$. Since $M_i'(x)$ and $m_i(x)$ are relatively prime, so are $m_i(x)$ and the following $M_i(x)$ defined by

$$M_i(x) = M_i'(x) \bmod m_i(x).$$

By the Euclidean algorithm we can find a pair of polynomials $a_i(x)$ and $b_i(x)$ such that

$$a_i(x)M_i(x) + b_i(x)m_i(x) = 1. \qquad (6.17)$$

By the Chinese Remainder Theorem for Polynomials we have

$$u(x) = \sum_{i=1}^{s} M_i'(x)[a_i(x)r_i(x) \bmod m_i(x)]. \qquad (6.18)$$

Thus, given s residues $r_i(x), i = 1, ..., s$, the remaining t residues $r_{s+1}(x)$, ..., $r_{s+t}(x)$ can be found by

$$r_{s+j}(x) = \sum_{i=1}^{s}[M_i'(x) \bmod m_{s+j}(x)][a_i(x)r_i(x) \bmod m_i(x)]. \qquad (6.19)$$

By (6.17) the polynomials $a_i(x)$ and $m_i(x)$ are also relatively prime. Define

$$h_{li}(x) = M_i'(x) \bmod m_{s+l}(x), \quad i = 1, ..., s; \; l = 1, ..., t. \qquad (6.20)$$

By definition $h_{li}(x)$ and $m_{s+l}(x)$ are relatively prime. Hence, the following two mappings from $GF(q)[x]_m$ to $GF(q)[x]_m$

$$\begin{aligned} \theta_i(r(x)) &= a_i(x)r(x) \bmod m_i(x), & i = 1, 2, ..., s, \\ \Theta_{li}(r(x)) &= h_{li}(x)r(x) \bmod m_{s+l}(x), & l = 1, 2, ..., t, \end{aligned} \qquad (6.21)$$

are one-to-one and linear with respect to the addition of $GF(q)[x]_m$.

Combining (6.19) and (6.21) yields

$$r_{s+l}(x) = \sum_{i=1}^{s} \Theta_{li}(\theta_i(r_i(x))) = \sum_{i=1}^{s} \Delta_{li}(r_i(x)), \qquad (6.22)$$

where Δ_{li} is the composite one-to-one linear transformation $\Theta_{li}\theta_i$. Under the natural one-to-one mapping a polynomial of $GF(q)[x]_m$ can be identified with its corresponding vector of $GF(q)^m$. Thus, the transformations Δ_{li} can be regarded as nonsingular linear transformations from $GF(q)^m$ to $GF(q)^m$, and can be represented by a matrix A_{li}. Set

$$A = \begin{bmatrix} A_{11} & A_{12} & \cdots & A_{1s} \\ A_{21} & A_{22} & \cdots & A_{2s} \\ \vdots & \vdots & \vdots & \vdots \\ A_{t1} & A_{t2} & \cdots & A_{ts} \end{bmatrix}. \qquad (6.23)$$

In terms of this matrix the code over $GF(q)$ can alternatively be described by

$$A \begin{bmatrix} \mathbf{r}_1 \\ \mathbf{r}_2 \\ \vdots \\ \mathbf{r}_s \end{bmatrix} = \begin{bmatrix} \mathbf{r}_{s+1} \\ \mathbf{r}_{s+2} \\ \vdots \\ \mathbf{r}_{s+t} \end{bmatrix}. \tag{6.24}$$

Before describing the decoding procedure we need the following lemma due to Bossen and Yau [14], whose proof is straightforward.

Lemma 6.5.1 *Any $w \times w$ submatrix of A given by (6.23) of the form*

$$\begin{bmatrix} A_{i_1 j_1} & \cdots & A_{i_1 j_w} \\ \vdots & \vdots & \vdots \\ A_{i_w j_1} & \cdots & A_{i_w j_w} \end{bmatrix}$$

is invertible, where $w \leq \min\{s, t\}$.

The characterization (6.24) and Lemma 6.5.1 play an important role in deriving the decoding procedure for the Bossen-Yau codes. With arguments similar to those for the majority logic decoding for the Reed-Solomon code over $GF(q^m)$ in Section 6.3, we know that the code $\varepsilon(\mathcal{C})$ can correct $t/2$ residues in error of a received word $(\mathbf{r}_1, ..., \mathbf{r}_{s+t})$ if t is even. The code $\mathcal{C}$ of (6.14) has ms information symbols and mt check symbols. Any burst of length $l \leq (t/2 - 1)m + 1$ symbols over $GF(q)$ can affect at most $t/2$ residues, and thus be corrected.

Based on the alternative characterization of (6.24), an alternative encoding process suggested by Bossen and Yau is the following. Let $\mathbf{r}_1, ..., \mathbf{r}_s$ be the ms information symbol. Then the mt redundant symbols $\mathbf{r}_{s+1}, ..., \mathbf{r}_{s+t}$ are calculated by (6.24). In what follows we shall consider the decoding process in the presence of burst errors.

Let $(\mathbf{r}_1, ..., \mathbf{r}_s, \mathbf{r}_{s+1}, ..., \mathbf{r}_{s+t})$ be the received codeword. The basic idea of the Bossen-Yau decoding process is to calculate the values for the redundant residues based on the nonredundant residues $\mathbf{r}_1, ..., \mathbf{r}_s$ in the received codeword. Let $\mathbf{r}'_{s+1}, ..., \mathbf{r}'_{s+t}$ be the t calculated redundant residues. The difference

$$(\mathbf{e}_1, ..., \mathbf{e}_t) = (\mathbf{r}'_{s+1} - \mathbf{r}_{s+1}, ..., \mathbf{r}'_{s+1} - \mathbf{r}_{s+1}) \tag{6.25}$$

corresponds to the syndrome. The relations among $\mathbf{e}_i$ and the matrices A_{ji} uniquely determine which burst of length $t/2$ or less occurred. To derive those relations we still need the following lemma.

Lemma 6.5.2 *Let* **v** *be a codeword of the Bossen-Yau code* $\varepsilon(\mathcal{C})$ *with the parameter* t *being even. Let* $\mathbf{v}' = \mathbf{v} + \mathbf{e}$, *where* $\mathrm{wt}(\mathbf{e}) \leq t/2$. *Then* $\mathbf{v}' - \mathbf{e}'$ *is a codeword of* $\varepsilon(\mathcal{C})$ *for some* $\mathbf{e}'$ *with* $\mathrm{wt}(\mathbf{e}') \leq t/2$ *if and only if* $\mathbf{e} = \mathbf{e}'$.

To correct a burst of length $t/2$ or less with the code $\varepsilon(\mathcal{C})$, we first calculate the syndrome as in (6.25). The syndrome is then fed into a set of logic circuits L_i, where the circuit L_i corrects a burst of $t/2$ or less residues in error which begins in position i for $i = 1, ..., s$. The output of the circuit L_i is a set $\mathbf{E}_i$, $\mathbf{E}_{i+1}$, ..., $\mathbf{E}_{i+t/2-1}$ such that

$$(\mathbf{r}_1, ..., \mathbf{r}_{s+t}) - (\mathbf{0}, ..., \mathbf{0}, \mathbf{E}_i, \mathbf{E}_{i+1}, ..., \mathbf{E}_{i+t/2-1}, \mathbf{0}, ..., \mathbf{0}) \in \varepsilon(\mathcal{C}).$$

We distinguish among the following three classes of errors.

Case 1: A burst of length no more than $t/2$ affects only the information residues $\mathbf{r}_1, ..., \mathbf{r}_s$.

Case 2: A burst of length no more than $t/2$ affects some of the information residues and some of the redundant residues $r_{s+1}, ..., r_{s+t}$.

Case 3: A burst of length no more than $t/2$ affects only the redundant residues.

In practice we are only interested in correcting the information residues, so it is necessary only to detect that Case 3 has occurred.

To characterize the burst errors, we define a set of matrices A_i for $i = 1, 2, ..., s$, as follows. For $1 \leq i \leq s - t/2 + 1$ we define

$$A_i = \begin{bmatrix} A_{1,i} & A_{1,i+1} & \cdots & A_{1,i+t/2-1} \\ A_{2,i} & A_{2,i+1} & \cdots & A_{2,i+t/2-1} \\ \vdots & \vdots & \vdots & \vdots \\ A_{t,i} & A_{t,i+1} & \cdots & A_{t,i+t/2-1} \end{bmatrix}. \tag{6.26}$$

For $s - t/2 + 2 \leq i \leq s$ we define

$$A_i = \begin{bmatrix} A_{1,i} & A_{1,i+1} & \cdots & A_{1,s} & & -I' \\ A_{2,i} & A_{2,i+1} & \cdots & A_{2,s} & & \\ \vdots & \vdots & \vdots & \vdots & & 0 \\ A_{t,i} & A_{t,i+1} & \cdots & A_{t,s} & & \end{bmatrix}, \tag{6.27}$$

where I' is the $m(t/2 - s - 1 + i) \times m(t/2 - s - 1 + i)$ identity matrix.

Let $(\mathbf{r}_1, ..., \mathbf{r}_{s+t})$ be a received codeword which is assumed to contain a burst of at most $t/2$ residues in error. Then the syndrome is described by

$$\mathbf{e}_j = \sum_{i=1}^{s} A_{ji}\mathbf{r}_i - \mathbf{r}_{s+j}. \tag{6.28}$$

Let the $(s+t)$-tuple $(\mathbf{0}, ..., \mathbf{0}, \mathbf{E}_1^i, ..., \mathbf{E}_{t/2}^i, \mathbf{0}, ..., \mathbf{0})$ represent a burst of length $\leq t/2$ beginning in position i, and define

$$\mathbf{E}^i = (\mathbf{E}_1^i, \mathbf{E}_2^i, ..., \mathbf{E}_{t/2}^i)^T.$$

For $1 \leq i \leq s - t/2 + 1$, define the s-tuple

$$\mathbf{E}_i = (\mathbf{0}, ..., \mathbf{0}, \mathbf{E}_1^i, ..., \mathbf{E}_{t/2}^i, \mathbf{0}, ..., \mathbf{0})^T.$$

For $s - t/2 + 2 \leq i \leq s$, define the s-tuple $\mathbf{E}_{i1}$ and t-tuple $\mathbf{E}_{i2}$ respectively as

$$\begin{aligned}
\mathbf{E}_{i1} &= (\mathbf{0}, ..., \mathbf{0}, \mathbf{E}_1^i, ..., \mathbf{E}_{s-i+1}^i)^T, \\
\mathbf{E}_{i2} &= (\mathbf{E}_{s-i+2}^i, ..., \mathbf{E}_{t/2}^i, \mathbf{0}, ..., \mathbf{0})^T.
\end{aligned}$$

For $s < i \leq s + t$, define the t-tuple

$$\mathbf{E}_i = (\mathbf{0}, ..., \mathbf{0}, \mathbf{E}_1^i, ..., \mathbf{E}_{t/2}^i, \mathbf{0}, ..., \mathbf{0})^T.$$

The following theorem is useful in characterizing the burst errors.

Theorem 6.5.3 *Let $(\mathbf{r}_1, ..., \mathbf{r}_{s+t})$ contain a burst of at most $t/2$ residues in error. Then*

$$\mathbf{v} = (\mathbf{r}_1, ..., \mathbf{r}_{s+t}) - (\mathbf{0}, ..., \mathbf{0}, (\mathbf{E}^i)^T, \mathbf{0}, ..., \mathbf{0})$$

is in $\varepsilon(\mathcal{C})$ if and only if

(i) $A_i \mathbf{E}^i = \mathbf{e}$ *for* $1 \leq i \leq s$, *where* $\mathbf{e} = (\mathbf{e}_1, ..., \mathbf{e}_t)^T$; *or*

(ii) $\mathrm{wt}(\mathbf{e}) \leq t/2$ *and* $\mathbf{E}_i = -\mathbf{e}$ *for* $s < i \leq s + t$.

Proof: Let

$$\begin{aligned}
\mathbf{V}_s &= (\mathbf{r}_1, ..., \mathbf{r}_s)^T, \\
\mathbf{V}_t &= (\mathbf{r}_{s+1}, ..., \mathbf{r}_{s+t})^T.
\end{aligned}$$

Suppose that $1 \leq i \leq s - t/2 + 1$ and $\mathbf{v}$ is in $\varepsilon(\mathcal{C})$. Then

$$A(\mathbf{V}_s - \mathbf{E}_i) - \mathbf{V}_t = \mathbf{0}$$

and hence

$$A\mathbf{E}_i = A\mathbf{V}_s - \mathbf{V}_t = \mathbf{e}. \tag{6.29}$$

By the definition of A_i and $\mathbf{E}_i$, equation (6.29) implies (i) for $1 \leq i \leq s - t/2 - 1$.

Assume now that $s - t/2 + 2 \leq i \leq s$ and $\mathbf{v}$ is in $\varepsilon(\mathcal{C})$. Then

$$A(\mathbf{V}_s - \mathbf{E}_{i1}) - \mathbf{V}_t + \mathbf{E}_{i2} = \mathbf{0}$$

and hence

$$A\mathbf{V}_s - \mathbf{V}_t = A\mathbf{E}_{i1} - \mathbf{E}_{i2} = \mathbf{e}. \tag{6.30}$$

By the definition of A_i and $\mathbf{E}^i$, equation (6.30) implies (i) for $s - t/2 + 2 \leq i \leq s$. It is easy to show that if (i) holds, then $\mathbf{v}$ is in $\varepsilon(\mathcal{C})$.

Suppose now that for $s < i \leq s + t$ we have $\mathbf{v} \in \varepsilon(\mathcal{C})$. Then $\mathbf{v}$ has syndrome $\mathbf{0}$ so that

$$A\mathbf{V}_s - \mathbf{V}_t + \mathbf{E}_i = \mathbf{0}$$

or $\mathbf{E}_i = -\mathbf{e}$, implying that $\text{wt}(\mathbf{e}) \leq t/2$. If (ii) holds, it is easy to verify that $\mathbf{v}$ has $\mathbf{0}$ syndrome and therefore is in $\varepsilon(\mathcal{C})$. □

Our task now is to find the conditions under which (i) or (ii) has a solution $\mathbf{E}^i$. If such a solution can be found, then the correct information symbols are obtained by subtracting the s-tuple $\mathbf{E}_i$ from $\mathbf{V}_s$.

Case 1: $1 \leq i \leq s - t/2 + 1$

Let $\mathbf{e}^1 = (\mathbf{e}_1, ..., \mathbf{e}_{t/2})^T$ and $\mathbf{e}^2 = (\mathbf{e}_{t/2+1}, ..., \mathbf{e}_t)^T$, where $\mathbf{e}_j$'s are defined as before. We now partition the matrix A_i into

$$A_i = \begin{bmatrix} B_{i1} \\ B_{i2} \end{bmatrix},$$

where $1 \leq i \leq s - t/2 + 1$ and

$$B_{i1} = \begin{bmatrix} A_{1,i} & A_{1,i+1} & \cdots & A_{1,i+t/2-1} \\ A_{2,i} & A_{2,i+1} & \cdots & A_{2,i+t/2-1} \\ \vdots & \vdots & \vdots & \vdots \\ A_{t/2,i} & A_{t/2,i+1} & \cdots & A_{t/2,i+t/2-1} \end{bmatrix},$$

$$B_{i2} = \begin{bmatrix} A_{t/2+1,i} & A_{t/2+1,i+1} & \cdots & A_{t/2+1,i+t/2-1} \\ A_{t/2+2,i} & A_{t/2+2,i+1} & \cdots & A_{t/2+2,i+t/2-1} \\ \vdots & \vdots & \vdots & \vdots \\ A_{t,i} & A_{t,i+1} & \cdots & A_{t,i+t/2-1} \end{bmatrix}.$$

By Lemma 6.5.2 the matrices B_{i1} and B_{i2} are invertible. Therefore, the matrix

$$B_i = \begin{bmatrix} B_{i1}^{-1} & 0 \\ B_{i2}B_{i1}^{-1} & -I \end{bmatrix}$$

has the inverse matrix

$$B_i^{-1} = \begin{bmatrix} B_{i1} & 0 \\ B_{i2} & -I \end{bmatrix},$$

where I and 0 are the identity and zero matrix of order $mt/2 \times mt/2$ respectively. Thus (i) has a solution if and only if

$$B_i A_i \mathbf{E}^i = B_i \mathbf{e} \tag{6.31}$$

has a solution $\mathbf{E}^i$. But (6.31) is equivalent to the following two equations

$$\begin{cases} \mathbf{E}^i = B_{i1}^{-1} \mathbf{e}^1, \\ 0 = B_{i2} B_{i1}^{-1} \mathbf{e}^1 - \mathbf{e}^2. \end{cases} \tag{6.32}$$

Thus the solution $\mathbf{E}^i$ in (6.32) exists if and only if

$$0 = B_{i1}^{-1} \mathbf{e}^1 - B_{i2}^{-1} \mathbf{e}^2. \tag{6.33}$$

The equations of (6.32) are the basis for the detection and correction of burst errors in Case 1, where a bust of no more than $t/2$ residues in error beginning in position i occurred, and $1 \le i \le s - t/2 + 1$. The above derivation is actually the process of solving the linear equation $A_i \mathbf{E}^i = \mathbf{e}$ for $1 \le i \le s - t/2 + 1$.

Case 2: $s - t/2 + 2 \le i \le s$

In this case a burst of length no more than $t/2$ beginning in position i occurred, where $s - t/2 + 2 \le i \le s$. Thus the burst error affects both the information and redundant residues. To solve equation (i) in this case, we partition the matrix A_i given by (6.27) into

$$A_i = \begin{bmatrix} C_i & -I' \\ & 0 \end{bmatrix},$$

where the matrix C_i is given by

$$C_i = \begin{bmatrix} A_{1,i} & A_{1,i+1} & \cdots & A_{1,s} \\ A_{2,i} & A_{2,i+1} & \cdots & A_{2,s} \\ \vdots & \vdots & \vdots & \vdots \\ A_{t,i} & A_{t,i+1} & \cdots & A_{t,s} \end{bmatrix}.$$

Let

$$\mathbf{E}'^i = (\mathbf{E}_{s-i+2}^i, ..., \mathbf{E}_{t/2}^i, \mathbf{E}_1^i, ..., \mathbf{E}_{s-i+1}^i)^T$$

and

$$A_i' = \begin{bmatrix} -I' & C_i \\ 0 & \end{bmatrix}.$$

Then $A_i \mathbf{E}^i = \mathbf{e}$ has a solution E^i if and only if

$$A_i' \mathbf{E}'^i = \mathbf{e} \tag{6.34}$$

has a solution $\mathbf{E}'^i$. Clearly, $\mathbf{E}^i$ can be obtained by a rearrangement of E'^i.
Now we partition the matrix C_i into

$$C_i = \begin{bmatrix} D_{i1} \\ D_{i2} \\ D_{i3} \\ D_{i4} \end{bmatrix}, \tag{6.35}$$

where D_{i1} and D_{i3} have dimension $m(t/2 - s - 1 + i) \times m(s - i + 1)$ and D_{i2} and D_{i4} have dimension $m(s - i + 1) \times m(s - i + 1)$. By Lemma 6.5.1 D_{i2} and D_{i4} are invertible. Then the following matrix

$$D_i = \begin{bmatrix} -I' & D_{i1}D_{i2}^{-1} & 0 \\ 0 & D_{i2}^{-1} & \\ 0 & D_{i3}D_{i2}^{-1} & \\ 0 & D_{i4}D_{i2}^{-1} & -I \end{bmatrix}$$

has the inverse

$$D_i^{-1} = \begin{bmatrix} -I' & D_{i1} & 0 \\ 0 & D_{i2} & \\ 0 & D_{i3} & \\ 0 & D_{i4} & -I \end{bmatrix},$$

where I is the $mt/2 \times mt/2$ identity matrix. Let

$$\begin{aligned} \mathbf{e}^{1i} &= (\mathbf{e}_1, ..., \mathbf{e}_{t/2-s+i-1})^T, \\ \mathbf{e}^{2i} &= (\mathbf{e}_{t/2-s+i}, ..., \mathbf{e}_{t/2})^T. \end{aligned}$$

Then (6.34) has a solution $\mathbf{E}'^i$ if and only if

$$D_i A_i' \mathbf{E}'^i = D_i \mathbf{e},$$

which is equivalent to

$$\mathbf{E}'^i = \begin{bmatrix} -\mathbf{e}^{1i} + D_{i1}D_{i2}^{-1}\mathbf{e}^{2i} \\ D_{i2}^{-1}\mathbf{e}^{2i} \end{bmatrix} \tag{6.36}$$

and

$$\mathbf{e}^2 = \begin{bmatrix} 0 & D_{i3}D_{i2}^{-1} \\ 0 & D_{i4}D_{i2}^{-1} \end{bmatrix} \mathbf{e}^1, \tag{6.37}$$

where the solution (6.36) exists if and only if (6.37) holds. Since we are only interested in correcting the information residues, we only need that portion of (6.36) given by

$$D_{i2}^{-1}\mathbf{e}^{2i}$$

which can be subtracted from the last $s-i+1$ residues of the received information residues to produce the correct residues.

Case 3: $s < i \leq s+t$

In this case a burst of length no more than $t/2$ occurred in the redundant residues only, and it is sufficient to detect that Case 3 has occurred, which can be done according to Theorem 6.5.3 by determining the number of nonzero elements in the syndrome. $t/2$ or fewer nonzero elements in the syndrome implies that Case 3 has occurred.

Summarizing the above analysis, we have the following theorem.

Theorem 6.5.4 *A burst of length no more than $t/2$ has occurred beginning in position i if and only if*

1. for $1 \leq i \leq s - t/2 + 1$

$$B_{i2}^{-1}\mathbf{e}^2 - B_{i1}^{-1}\mathbf{e}^1 = 0 \tag{6.38}$$

in which case the error is given by

$$B_{i1}^{-1}\mathbf{e}^1, \tag{6.39}$$

which should be subtracted from the received word beginning in position i to produce the correct word;

2. for $s - t/2 + 2 \leq i \leq s$

$$\mathbf{e}^2 = \begin{bmatrix} 0 & D_{i3}D_{i2}^{-1} \\ 0 & D_{i4}D_{i2}^{-1} \end{bmatrix} \mathbf{e}^1 \tag{6.40}$$

in which case the error is given by

$$D_{i2}^{-1}\mathbf{e}^{2i} \tag{6.41}$$

which should be subtracted from the received word beginning in position i to produce the correct codeword;

3. for $i > s$, the syndrome contains $t/2$ or fewer nonzero elements, in which case the information residues are correct.

To mechanize the above burst-error correcting process with special hardware, field multipliers and adders for $GF(q)$ are needed. But for the binary case, only modulo-2 adders are needed. To do this, Bossen and Yau have analyzed the circuitry required:

1) Generation of the syndrome defined by (6.28) requires ms modulo-2 adders.

2) For $1 \leq i \leq s - t/2 + 1$, the generation of (6.38) requires $mt/2$ modulo-2 adders plus a circuit to test for all 0's. The generation of the error of (6.39) requires an additional $mt/2$ modulo-2 adders. Thus, 2) requires a total of $mt(s - t/2 + 1)$ modulo-2 adders. Each modulo-2 adder requires at most $mt/2$ inputs.

3) For $s - t/2 + 2 \leq i \leq s$, the generation of (6.40) requires $mt/2$ modulo-2 adders. The generation of the error of (6.41) requires $m(s-i+1)$ modulo-2 adders. Thus, 3) requires a total of $mt(t/2 - 1) - m \sum_{j=1}^{t/2-1} j$ modulo-2 adders.

Therefore, 1), 2) and 3) require a total of

$$ms(t + 1) - \frac{mt}{4} \left(\frac{t}{2} - 1 \right)$$

modulo-2 adders.

The decoder developed by Bossen and Yau for the binary case is shown in Figure 6.1. The amount of hardware required is comparable with that required for the instantaneous decoding of existing burst-error correcting codes, such as the Fire codes. With a general-purpose computer, a software decoding algorithm is easy to implement. Since the code $\varepsilon(C)$ is MDS, the decoding procedure described by Mandelbaum [52], which is also based on the CRT, may be interesting to the decoding of Bossen-Yau codes.

6.6 Generalized Redundant Residue Codes

In this section we generalize redundant residue codes into a more general class, and show that Goppa codes belong to this class. This generalization is in fact based on the generalized CRT in Section 2.5. Decoding those codes is essentially the same as decoding the corresponding redundant residue codes. The following lemma is essential to the generalization of redundant residue codes.

Lemma 6.6.1 *Let F be a field, and $m_1(x), ..., m_s(x)$ be s pairwise relatively prime polynomials of $F[x]$. For any s nonzero elements $b_1, ..., b_s$ of F, the mapping*

$$\lambda : \begin{cases} F[x] & \rightarrow & F[x]/(m_1(x)) \times \cdots \times F[x]/(m_s(x)), \\ f(x) & \mapsto & (b_1 f_1(x), ..., b_s f_s(x)) \end{cases} \qquad (6.42)$$

is a group homomorphism with respect to the additions of the ring $F[x]$ and the residue class rings $F[x]/(m_i(x))$, where $f_i(x) = f(x) \bmod m_i(x)$ for $i = 1, ..., s$. Furthermore,

$$F[x]/(m(x)) \cong F[x]/(m_1(x)) \times \cdots \times F[x]/(m_s(x)), \qquad (6.43)$$

where the isomorphism is with respect to the additions of the rings, and $m(x) = \prod_{i=1}^{s} m_i(x)$.

It should be noted that λ is a ring homomorphism if and only if $b_1 = b_2 = \cdots = b_s = 1$, in which case we have the Chinese Remainder Theorem.

Let $m_0(x), m_1(x), \cdots, m_{s+t-1}(x) \in GF(q)[x]$ be pairwise relatively prime, with degrees $e_0, e_1, \cdots, e_{s+t-1}$ respectively, and $k = \sum_{i=0}^{s-1} e_i$. As before, for each polynomial $p(x)$ of degree no more than $k - 1$ let

$$r_i(x) = p(x) \bmod m_i(x), \quad i = 0, \cdots, s + t - 1.$$

Let $\mathbf{r}_i = (r_{i,0}, r_{i,1}, ..., r_{i,e_i-1})$ which corresponds to the residue $r_i(x) = r_{i,0} + r_{i,1}x + \cdots + r_{i,e_i-1}x^{e_i-1}$. The generalized redundant residue codes (briefly, GRR codes) are described by

$$C = \{(b_0 \mathbf{r}_0, \cdots, b_{s-1}\mathbf{r}_{s-1}, b_s \mathbf{r}_s, \cdots, b_{s+t-1}\mathbf{r}_{s+t-1}) | p(x) \in GF(q)[x]_k\}, \quad (6.44)$$

where $GF(q)[x]_k$ is the set of all polynomials of degree no more than $k - 1$ over $GF(q)$, and $b_0, ..., b_{s+t-1}$ are a set of nonzero elements of $GF(q)$.

Theorem 6.6.2 *For each set of nonzero elements $b_0, ..., b_{s+t-1}$ of $GF(q)$, the generalized redundant residue code given by (6.44) is an $[\sum_{i=0}^{s+t-1} e_i, k]$ linear code with the same minimum distance as the code given by (6.1), where $k = \sum_{i=0}^{s-1} e_i$.*

Proof: It follows from Lemma 6.6.1 and Theorem 6.2.1 that it is an $[\sum_{i=0}^{s+t-1} e_i, k]$ linear code. Since $b_0, ..., b_{s+t-1}$ are nonzero, the transform defined by this set of nonzero elements does not change the minimum distance. □

With a redundant residue code given by (6.1) we can get $(q-1)^{s+t-1}$ generalized redundant residue codes over $GF(q)$ having the same parameters as the original code of (6.1). When $b_0 = \cdots = b_{s+t-1} = 1$, the code of (6.44) coincides with that of (6.1). Obviously, the generalization is meaningful only when $q > 2$.

The encoding of the generalized redundant code described by (6.44) can be carried out similarly. Let $\mathbf{u} = (u_0, ..., u_{k-1})$ be an information vector. First, we compute the residues $r_i(x) = \sum_{i=0}^{k-1} u_i x^i \bmod m_i(x)$. Then the corresponding codeword is given by

$$\mathbf{c} = (b_0 \mathbf{r}_0, ..., b_{s+t-1} \mathbf{r}_{s+t-1}).$$

A slightly modified version of any decoding algorithm for the redundant residue code of (6.1) applies to the generalized redundant residue code of (6.44).

Now we describe the generalized redundant residue codes of some specific redundant residue codes. Let $m_i = x - \alpha^i$ for $0 \le i \le q - 2$, where α is a generating element of $GF(q)$, and $h_1(x), ..., h_t(x)$ be any t monic irreducible polynomials of degree two over $GF(q)$, where $1 \le t \le q(q-1)/2$. Then the polynomials $m_i(x)$ for $i = 0, ..., q-2$ and $h_1(x), ..., h_t(x)$ are pairwise relatively prime. The generalized redundant residue code of the code given by (6.9) is described by

$$\mathcal{Q}_1 = \{(b_0 \mathbf{r}_0, ..., b_{q-2} \mathbf{r}_{q-2}; b_{q-1} \mathbf{s}_1, ..., b_{q-2+t} \mathbf{s}_t) | u(x) \in GF(q)[x]_k\}, \qquad (6.45)$$

where $r_i(x) = u(x) \bmod m_i$ for $i = 0, ..., q - 2$, $s_i(x) = u(x) \bmod h_i(x)$, $\mathbf{s}_i$ is a vector of two components corresponding to the quadratic polynomial $s_i(x)$, where $1 \le i \le t$, and $b_0, ..., b_{q-2+t}$ are a set of nonzero elements of $GF(q)$.

Theorem 6.6.3 For any set of nonzero elements $b_0, ..., b_{q-2+t}$ of $GF(q)$, the generalized redundant residue code given by (6.45) is an $[q-1+2t, k, d \ge q-k+t]$ linear code over $GF(q)$.

Proof: The conclusion follows from Theorems 6.6.2 and 6.4.2. □

A very important special case of the generalized redundant residue code of (6.45) is $t = 1$. In this case it is a $[q+1, k, d]$ linear code for any set of nonzero elements $b_0, ..., b_q$, where $d = q - k + 1$ or $q - k + 2$ by Theorem 6.4.3.

Let $m_{q-1}(x) = x$ and $r_{q-1}(x) = u(x) \bmod m_{q-1}(x)$. The generalized redundant residue code of the code given by (6.13) is given by

$$\mathcal{Q}'_1 = \{(b_0 \mathbf{r}_0, ..., b_{q-2} \mathbf{r}_{q-2}, b_{q-1} \mathbf{r}_{q-1}; b_q \mathbf{s}_1, ..., b_{q+t-1} \mathbf{s}_t) | u(x) \in GF(q)[x]_k\}, (6.46)$$

where $b_0, ..., b_{q+t-1}$ are fixed nonzero elements of $GF(q)$, and the other symbols are the same as before.

Theorem 6.6.4 For any set of nonzero elements $b_0, ..., b_{q+t-1}$ of $GF(q)$, the generalized redundant residue code given by (6.46) is an $[q+2t, k, d \ge q-k+t+1]$ linear code over $GF(q)$.

Proof: The conclusion follows from Theorems 6.6.2 and 6.4.4.   □

An very interesting special case of the generalized redundant residue code of (6.46) is $t = 1$, in which case it is a $[q + 2, k, d \geq q - k + 2]$ linear code over $GF(q)$ corresponding to the code $\mathcal{R}'$ of Theorem 6.4.5.

An important class of generalized redundant residue codes is the generalized Reed-Solomon codes. Let $\Delta = (\delta_1, ..., \delta_n)$, where the δ_i are distinct elements of the field $GF(q^m)$, and $n \leq q^m$. And let $\mathbf{b} = (b_1, ..., b_n)$, where the b_i are nonzero (but not necessarily distinct) elements of $GF(q^m)$. As before, the moduli are chosen to be $m_i(x) = x - \delta_i$ for $i = 1, ..., n$. Recall that the generalized Reed-Solomon code over $GF(q^m)$, denoted by $\mathrm{GRS}_k(\Delta, \mathbf{b})$, is described by

$$\mathrm{GRS}_k(\Delta, \mathbf{b}) = \{(b_1 r_1, b_2 r_2, ..., b_n r_n) | u(x) \in GF(q^m)[x]_k\}, \qquad (6.47)$$

where $r_i = u(x) \bmod m_i$ for $i = 1, ..., n$ [51]. Since the moduli $m_i(x)$ are of degree one, the generalized Reed-Solomon code can also be described by

$$\mathrm{GRS}_k(\Delta, \mathbf{b}) = \{(b_1 u(\delta_1), b_2 u(\delta_2), ..., b_n u(\delta_n) | u(x) \in GF(q^m)[x]_k\}. \qquad (6.48)$$

Theorem 6.6.5 *The code $GRS_k(\Delta, \mathbf{b})$ is an $[n, k, n - k + 1]$ linear code, and the dual of $GRS_k(\Delta, \mathbf{b})$ is $GRS_{n-k}(\Delta, \mathbf{b}')$ for some $\mathbf{b}'$.*

Proof: As a special class of generalized redundant residue codes, the codes $\mathrm{GRS}_k(\Delta, \mathbf{b})$ must be linear. Note that a polynomial of degree less than k over $GF(q^m)$ has at most $k - 1$ zeros, δ_i are distinct, and b_i are nonzero. Hence, the minimum distance of the code must be $n - (k - 1) = n - k + 1$.

To prove the remaining conclusions, let $\mathbf{y} = (y_1, ..., y_n) \in GF(q^m)^n$ and

$$V = \begin{bmatrix} 1 & 1 & \cdots & 1 \\ \delta_1 & \delta_2 & \cdots & \delta_n \\ \delta_1^2 & \delta_2^2 & \cdots & \delta_n^2 \\ \vdots & \vdots & \vdots & \vdots \\ \delta_1^{n-2} & \delta_2^{n-2} & \cdots & \delta_n^{n-2} \end{bmatrix}.$$

Since δ_i are distinct and deleting any column of V yields a Vandermonde matrix of dimension $(n - 1) \times (n - 1)$, the rank of V is $n - 1$. Thus, the solution space of the following equation

$$V\mathbf{y}^T = \mathbf{0} \qquad (6.49)$$

has dimension one, and every solution of this equation can be expressed as

$$\mathbf{y} = a \cdot (d_1, d_2, ..., d_n) = (ad_1, ad_2, ..., ad_n),$$

where $a \in GF(q^m)$, $d_0, d_1, ..., d_n$ are nonzero elements of $GF(q^m)$, and $\mathbf{d} = (d_1, ..., d_n)$ is a solution of (6.49).

Now let $b_i' = d_i/b_i$ for $i = 1, 2, ..., n$, and consider the code $\mathrm{GRS}_{n-k}(\Delta, \mathbf{b}')$. Take any codeword $\mathbf{c} = (b_1 u(\delta_1), ..., b_n u(\delta_n))$ from $\mathrm{GRS}_k(\Delta, \mathbf{b})$ and any codeword $\mathbf{c}' = (b_1' v(\delta_1), ..., b_n' v(\delta_n))$ from $\mathrm{GRS}_{n-k}(\Delta, \mathbf{b}')$, where

$$u(x) = \sum_{i=0}^{k-1} u_i x^i, \quad v(x) = \sum_{i=0}^{n-k-1} v_i x^i.$$

The inner product of the two vectors is

$$
\begin{aligned}
\mathbf{c} \cdot \mathbf{c}' &= \sum_{j=0}^{k-1} u_j \left[\sum_{l=0}^{n-k-1} v_l \left[\sum_{i=1}^{n} \delta_i^{j+l} b_i b_i' \right] \right] \\
&= \sum_{j=0}^{k-1} u_j \left[\sum_{l=0}^{n-k-1} v_l \left[\sum_{i=1}^{n} \delta_i^{j+l} d_i \right] \right] \\
&= 0.
\end{aligned}
$$

Thus, the dual code of $\mathrm{GRS}_k(\Delta, \mathbf{b})$ contains the code $\mathrm{GRS}_{n-k}(\Delta, \mathbf{b}')$. But the dimension of $\mathrm{GRS}_{n-k}(\Delta, \mathbf{b}')$ is $n-k$, thus the dual code must be $\mathrm{GRS}_{n-k}(\Delta, \mathbf{b}')$. This completes the proof. In addition, the choice for the vector $\mathbf{b}'$ is not unique, but the resulting code $\mathrm{GRS}_{n-k}(\Delta, \mathbf{b}')$ is the same. The proof here is in fact a constructive one, and the actual dual code has been found. $\qquad \square$

From the proof of the above theorem it is easily seen that the parity check matrix of $\mathrm{GRS}_k(\Delta, \mathbf{b})$ is the generator matrix of $\mathrm{GRS}_{n-k}(\Delta, \mathbf{b}')$ which is

$$
M = \begin{bmatrix}
b_1' & b_2' & \cdots & b_n' \\
\delta_1 b_1' & \delta_2 b_2' & \cdots & \delta_n b_n' \\
\vdots & \vdots & \vdots & \vdots \\
\delta_1^{n-k-1} b_1' & \delta_2^{n-k-1} b_2' & \cdots & \delta_n^{n-k-1} b_n'
\end{bmatrix}.
$$

Though the class of GRR codes described by (6.44) contains many linear codes, it can be further extended by the generalized CRT for polynomials described by Theorem 2.5.3 as follows.

Let $m_0(x), m_1(x), \cdots, m_{s+t-1}(x) \in GF(q)[x]$ be pairwise relatively prime, with degrees $e_0, e_1, \cdots, e_{s+t-1}$ respectively, and $k = \sum_{i=0}^{s-1} e_i$. Choose polynomials $a_i(x) \in GF(q)[x]$ such that $\gcd(m_i(x), a_i(x)) = 1$ for $i = 0, 1, ..., s+t-1$. As before, for each polynomial $p(x)$ of degree no more than $k - 1$ let

$$r_i(x) = a_i(x) p(x) \bmod m_i(x), \quad i = 0, \cdots, s+t-1.$$

Let $\mathbf{r}_i = (r_{i,0}, r_{i,1}, ..., r_{i,e_i-1})$ which corresponds to the residue $r_i(x) = r_{i,0} + r_{i,1} x + \cdots + r_{i,e_i-1} x^{e_i-1}$. The *extended redundant residue codes* (briefly, ERR codes) are described by

$$\mathcal{C} = \{(\mathbf{r}_0, \cdots, \mathbf{r}_{s-1}, \mathbf{r}_s, \cdots, \mathbf{r}_{s+t-1}) | p(x) \in GF(q)[x]_k\}. \qquad (6.50)$$

This is an $[\sum_{i=0}^{s+t-1} e_i, k]$ linear code, but its minimum distance is usually not the same as that of the code given by (6.1). If $a_0(x) = a_1(x) = \cdots = a_{s+t-1}(x) = 1$, it is the code of (6.1). If $a_i(x) = b_i$ are nonzero elements of $GF(q)$, the code given by (6.50) is identical to the one given by (6.44). Properties of the ERR codes remain to be investigated, but decoding ERR codes is essentially the same as decoding GRR codes.

6.7   Restricted GRR Codes

In this section we shall introduce the family of restricted GRR codes and show that many well-known classes of linear codes belong to this family. Among them are alternant codes, Goppa codes, BCH codes, Chien-Choy generalized BCH codes, Generalized Srivastava codes. Since these codes have intensively been studied, we shall only show the importance of the generalized redundant residue codes without going deeply into special subfield codes of the family of GRR codes.

In the foregoing section we saw that generalized Reed-Solomon codes belong to the family of GRR codes. Let $m_0(x), m_1(x), ..., m_{s+t-1}(x)$ be pairwise relatively prime polynomials over $GF(q^m)$, with degrees $e_0, ..., e_{s+t-1}$ respectively, $n = \sum_{i=0}^{s+t-1} e_i$, and $k = \sum_{i=0}^{s-1} e_i$. Recall that the generalized redundant residue code over $GF(q^m)$ with respect to the set of nonzero elements $b_0, ..., b_{s+t-1}$ of $GF(q^m)$ is described by

$$\mathcal{C} = \{(b_0\mathbf{r}_0, ..., b_{s+t-1}\mathbf{r}_{s+t-1}) | u(x) \in GF(q^m)[x]_k\}, \qquad (6.51)$$

where $r_i(x) = u(x) \bmod m_i(x)$, and $\mathbf{r}_i$ is the vector corresponding to the polynomial $r_i(x)$ with $0 \le i \le s + t - 1$. Let $\mathbf{e} = (e_0, ..., e_{s+t-1})$, $\mathbf{b} = (b_0, ..., b_{s+t-1})$, and $M = \{m_0(x), ..., m_{s+t-1}\}$. For simplicity we use $\mathrm{GRR}_k(M, \mathbf{e}, \mathbf{b})$ to denote the code over $GF(q^m)$ defined by (6.51).

The restricted GRR code, denoted by $\mathrm{RGRR}_k(M, \mathbf{e}, \mathbf{b})$, consists of all codewords of the $\mathrm{GRR}_k(M, \mathbf{e}, \mathbf{b})$ which have components from $GF(q)$, i.e., $\mathrm{RGRR}_k(M, \mathbf{e}, \mathbf{b})$ is the restriction of $\mathrm{GRR}_k(M, \mathbf{e}, \mathbf{b})$ to $GF(q)$.

Theorem 6.7.1 *The code* $\mathrm{RGRR}_k(M, \mathbf{e}, \mathbf{b})$ *is an* $[n, k', d']$ *linear code over* $GF(q)$ *with* $d' \ge d$, *the minimum distance of the* $[n, k, d]$ *code* $\mathrm{GRR}_k(M, \mathbf{e}, \mathbf{b})$, *and*

$$n - m(n - k) \le k' \le k,$$

where n, m, *and* k *are defined as before.*

Proof: Since $\mathrm{RGRR}_k(M, \mathbf{e}, \mathbf{b})$ is a subfield code of $\mathrm{GRR}_k(M, \mathbf{e}, \mathbf{b})$, the conclusions follow. □

A class of RGRR codes is the *alternant codes*, which are defined to be the restriction of $[n, k, n - k + 1]$ codes $\mathrm{GRS}_k(\Delta, \mathbf{b})$ over $GF(q^m)$ to $GF(q)$. By Theorem 6.7.1 they are $[n, n - m(n - k) \leq k' \leq k, d \geq n - k + 1]$ linear codes over $GF(q)$.

A subclass of the alternant codes is the Goppa codes, which are special RGRR codes. Let the set $\Delta = \{\delta_1, ..., \delta_n\}$ be a set of distinct elements of $GF(q^m)$, and $G(x)$ be a polynomial of $GF(q^m)[x]$ such that $G(\delta_i) \neq 0$ for all $\delta_i \in \Delta$. Then the Goppa code $\Gamma(\Delta, G)$ is defined to be the restriction of $\mathrm{GRS}_{n-k}(\Delta, \mathbf{b})$, where $\mathbf{b} = (b_1, ..., b_n)$ with

$$b_i = \frac{G(\delta_i)}{\prod_{j \neq i}(\delta_i - \delta_j)}, \quad i = 1, ..., n.$$

There are also other special alternant codes, such as the Srivastava codes and generalized Srivastava codes, BCH and Chien-Choy generalized BCH codes, which belong to the family of RGRR codes. For details about these codes we refer to [51, Chapter 12].

We have seen that RGRR codes form a large family of codes. Some of those codes have been studied intensively, and some of their parameters are known; but many of them have not been attacked yet. Thus, further investigations into the family of RGRR codes are needed. Similarly, we have the restricted ERR codes, which should contain more good codes.

6.8 A Class of Arithmetic Residue Codes

We first recall the idea of constructing redundant polynomial residue codes in the preceding sections, where the Chinese Remainder Theorem for polynomials is employed. Applying the same idea and the CRT for integers gives a class of arithmetic residue codes which can be described as follows [54].

Choose $n + r$ integer moduli m_i such that $\gcd(m_i, m_j) = 1$ for each pair of i and j with $i \neq j$, and $m_i > m_j$ if $i > j$, where $1 \leq i \leq n + r$ and $1 \leq j \leq n + r$. Let

$$M_n = \prod_{i=1}^{n} m_i, \quad R = \prod_{i=n+1}^{n+r} m_i, \quad M_r = RM_n.$$

The information number x, which is a number in the range $0 \leq x \leq M_n - 1$, is encoded to be an $(n + r)$-tuple given by

$$\mathbf{x} = (x_1, ..., x_{n+r}) \in \mathbf{Z}/(m_1) \times \cdots \times \mathbf{Z}/(m_{n+r}),$$

where $x_i = x \bmod m_i$ for $i = 1, ..., n + r$. By the CRT and the choice of the moduli, any n components of $\mathbf{x} = (x_1, ..., x_{n+r})$ suffice to recover the original information number x. The first n residues $x_1, ..., x_n$ are information residues, while the last r residues are redundant residues for error-detection and -correction.

An arithmetic code $\mathcal{C}$ is a subset of $\mathbf{Z}/(m_1) \times \cdots \times \mathbf{Z}/(m_{n+r})$. For two code vectors $\mathbf{u} = (u_1, ..., u_{n+r}), \mathbf{v} = (v_1, ..., v_{n+r}) \in \mathbf{Z}/(m_1) \times \cdots \times \mathbf{Z}/(m_{n+r})$, the Hamming distance between $\mathbf{u}$ and $\mathbf{v}$ is defined to be the number of nonzero components in $(u_1 - v_1, ..., u_{n+r} - v_{n+r})$, and the Hamming weight of $\mathbf{u}$ is the number of nonzero components in $\mathbf{u}$. By t errors in a received vector $\mathbf{y} = (y_1, ..., y_{n+r})$ we mean that exact t components of $\mathbf{y} = (y_1, ..., y_{n+r})$ are different from those of the transmitted vector. The Hamming distance of an arithmetic code is the minimum distance between all pairs of distinct code vectors in $\mathcal{C}$.

Let the symbols be the same as before, and define the arithmetic code by

$$\mathcal{C} = \{\mathbf{x} = (x_1, ..., x_{n+r}) \in \mathbf{Z}/(m_1) \times \cdots \times \mathbf{Z}/(m_{n+r})|x = 0, 1, ..., M_n - 1\}(6.52)$$

where $x_i = x \bmod m_i$ for $i = 1, ..., n + r$. This code is usually not linear with respect to the additions of the residue class rings $\mathbf{Z}/(m_i)$.

We now observe an example. Choose $n = 2, r = 2, m_1 = 3, m_2 = 4, m_3 = 5, m_4 = 7$. These moduli are pairwise relatively prime and $M_n = 12, R = 35$, and $M_r = 12 \times 35$. The information number 10 is encoded as $(1, 2, 0, 3)$ and the corresponding arithmetic code is given by

$$\mathcal{C}_1 = \left\{ \begin{array}{llll} (0,0,0,0), & (1,1,1,1), & (2,2,2,2), & (0,3,3,3), \\ (1,0,4,4), & (2,1,0,5), & (0,2,1,6), & (1,3,2,0), \\ (2,0,3,1), & (0,1,4,2), & (1,2,0,3), & (2,3,1,4) \end{array} \right\}.$$

It is easily checked that both of the minimum nonzero Hamming weight and the minimum distance of this arithmetic code are three. Generally, we have the following conclusion.

Theorem 6.8.1 *The arithmetic code given by (6.52) has the minimum nonzero Hamming weight $w_{min} \geq r + 1$ and minimum distance $d \geq r + 1$.*

Proof: By definitions the product of any n moduli among $\{m_1, ..., m_{n+r}\}$ is greater than or equal to M_n. Thus any n residues of $\{x_1, ..., x_{n+r}\}$ are sufficient to determine the information number x by the CRT. If the Hamming weight of $\mathbf{x} = (x_1, ..., x_{n+r})$ is less than $r + 1$, there must exist at least $(n + r - r)$ zero components in the vector $\mathbf{x}$. It follows from the CRT that $x = 0$. Hence, the Hamming weight of any nonzero code vector of $\mathcal{C}$ is no less than $r + 1$.

Let $\mathbf{x} = (x_1, ..., x_{n+r})$ and $\mathbf{y} = (y_1, ..., y_{n+r})$ be two code vectors of $\mathcal{C}$. If the distance between $\mathbf{x}$ and $\mathbf{y}$ is less than $r+1$, the Hamming weight of $\mathbf{x} - \mathbf{y} = (x_1 -$

$y_1, ..., x_{n+r} - y_{n+r})$ is less than $r+1$. Then there are at least n zero components $x_i - y_i$. By the CRT we have $x - y = 0$ mod M_n. Since $0 \leq x \leq M_n - 1$ and $0 \leq x \leq M_n - 1$, we have $x = y$. This completes the proof. $\square$

Theorem 6.8.2 *The arithmetic residue code given by (6.52) can detect all t errors for $t \leq r$, and correct $\lfloor r/2 \rfloor$ or fewer errors.*

Proof: The conclusions follow easily from Theorem 6.8.1. $\square$

The above arithmetic code given by (6.52) can be generalized as follows. Let the symbols be the same as before, and choose a set of positive integers $A = \{a_1, ..., a_{n+r}\}$ such that $\gcd(a_i, m_i) = 1$ for $i = 1, ..., n+r$. The generalized arithmetic code of (6.52) is given by

$$
\begin{aligned}
\mathcal{C}_A &= \{(a_1 x_1, ..., a_{n+r} x_{n+r}) \in \\
&= \mathbf{Z}/(m_1) \times \cdots \times \mathbf{Z}/(m_{n+r}) | x = 0, 1, ..., M_n - 1\},
\end{aligned}
\tag{6.53}
$$

where $x_i = x$ mod m_i for $i = 1, ..., n+r$. The encoding of this code is the same as the one given by (6.52). A slight modification of the decoding procedure outlined below for the code of (6.52) gives an efficient one for this generalized arithmetic code.

Just as for the redundant polynomial residue codes described in the preceding sections, all of the encoding, decoding and parameter analysis of the arithmetic residue codes given in this section are based on the CRT and CRA. An efficient decoding algorithm for those arithmetic residue codes was first suggested by Mandelbaum [54], and was improved by Barsi and Maestrini [7] and by Mandelbaum [56] respectively. The algorithm is based on the CRA and the continued fraction algorithm.

Recall now the CRA. Let the symbols be the same as before. Let $m_i' = M_r/m_i$ for $i = 1, ..., n+r$. Then $\gcd(m_i', m_i) = 1$. By the Euclidean algorithm we can find two integers p_i and q_i such that

$$
m_i' p_i + m_i q_i = 1.
$$

For $\mathbf{x} = (x_1, ..., x_{n+r})$ the CRA gives the original x by

$$
\begin{aligned}
x &= \sum_{i=1}^{n+r} m_i' p_i x_i \text{ mod } M_r \\
&= \sum_{i=1}^{n+r} \frac{M_r(p_i x_i \text{mod} m_i)}{m_i} \text{ mod } M_r.
\end{aligned}
\tag{6.54}
$$

Let $\mathbf{x} = (x_1, ..., x_{n+r})$ be the transmitted vector and $\mathbf{x}' = (x_1', ..., x_{n+r}')$ the received vector. If there are no errors in $\mathbf{x}'$, the information number is given by (6.54). Suppose $r = 2t$ and that t errors occurred in $\mathbf{x}'$, the received vector is of the form

$$
\mathbf{x}' = \mathbf{x} + (0, ..., e_{i_1}, 0, ..., 0, e_{i_t}, 0, ..., 0),
$$

where the e_{i_j} indicate errors. By the CRA the received number is

$$
\begin{aligned}
x' &= x + \sum_{j=1}^{t} \frac{M_r(p_{i_j} e_{i_j} \bmod m_{i_j})}{m_{i_j}} \bmod M_r \\
&= x + M_r \frac{B}{C} \bmod M_r,
\end{aligned}
\tag{6.55}
$$

where $C = \prod_{j=1}^{t} m_{i_j}$.

Note that M_r/C is the product of $n + t$ moduli, we have

$$
M_n = \prod_{i=1}^{n} m_i < \prod_{i=1}^{n+t} m_i \leq \frac{M_r}{C}
$$

and

$$
M_n < \frac{M_r B}{C} \bmod M_r < M_r.
$$

Therefore $M_r B/C \bmod M_r$ can be written as $M_r B'/C$; where $B' = B \bmod C$. It is easily seen that

$$
\frac{1}{R} < \frac{B'}{C} < 1 - \frac{1}{R}.
$$

Hence we have

$$
x' = x + M_r B'/C.
$$

Thus the decoding problem is to find $M_r B'/C$, and therefore B'/C by information about B'/C. Since $0 \leq x \leq M_n - 1$, it holds

$$
x' - x = M_r B'/C \leq x',
\tag{6.56}
$$

where x is unknown. Thus if we can find B'/C such that C is a product of t or fewer distinct moduli where $1/R < B'/C < 1 - (1/R)$ and (6.56) holds (given x'), then $M_r B'/C$ is the correct error term and B'/C is the unique error fraction. Thus $x = x' - M_r B'/C$ is the correct code number.

Before introducing the decoding procedure, we need a number of results about continued fractions. An expression of the form

$$
a_0 + \cfrac{1}{a_1 + \cfrac{1}{a_2 + \cdots + \cfrac{1}{a_N}}}
$$

is called a finite continued fraction and is usually written as

$$
a_0 + \frac{1}{a_1 +} \frac{1}{a_2 +} \cdots \frac{1}{a_N}
$$

or as

$$[a_0, a_1, a_2, \cdots, a_N].$$

The $a_0, ..., a_N$ are called *partial quotients* or simply *quotients* of the continued fraction.

By calculation we have

$$[a_0] = \frac{a_0}{1}, \quad [a_0, a_1] = \frac{a_1 a_0 + 1}{a_1}, \quad [a_0, a_1, a_2] = \frac{a_2 a_1 a_0 + a_2 + a_0}{a_2 a_1 + 1}$$

and

$$[a_0, a_1, ..., a_{n-1}, a_n] = [a_0, a_1, ..., a_{n-2}, a_{n-1} + \frac{1}{a_n}],$$

$$[a_0, a_1, ..., a_n] = a_0 + \frac{1}{[a_1, a_2, ..., a_n]} = [a_0, [a_1, a_2, ..., a_n]]$$

for $1 \le n \le N$. More generally

$$[a_0, a_1, ..., a_n] = [a_0, ..., a_{m-1}, [a_m, ..., a_n]]$$

for $1 \le m < n \le N$. The $[a_0, a_1, ..., a_n]$ is called the nth *convergent* to $[a_0, a_1, ..., a_N]$. By induction it is easy to prove the following theorem.

Theorem 6.8.3 *If p_n and q_n are defined by*

$$\begin{aligned} p_0 = a_0, \quad & p_1 = a_1 a_0 + 1, \quad p_n = a_n p_{n-1} + p_{n-2} \quad (2 \le n \le N), \\ q_0 = 1, \quad & q_1 = a_1, \quad\quad\quad\; q_n = a_n q_{n-1} + q_{n-2} \quad (2 \le n \le N), \end{aligned} \qquad (6.57)$$

then

$$[a_0, a_1, ..., a_n] = \frac{p_n}{q_n}. \qquad (6.58)$$

This theorem gives an efficient method to calculate the convergent. A useful result for our application is the following theorem whose proof is straightforward.

Theorem 6.8.4 *Let the symbols be the same as before. Then*

$$p_n q_{n-1} - p_{n-1} q_n = (-1)^{n-1}$$

or

$$\frac{p_n}{q_n} - \frac{p_{n-1}}{q_{n-1}} = \frac{(-1)^{n-1}}{q_{n-1} q_n}$$

and

$$p_n q_{n-2} - p_{n-2} q_n = (-1)^n a_n$$

or

$$\frac{p_n}{q_n} - \frac{p_{n-2}}{q_{n-2}} = \frac{(-1)^n a_n}{q_{n-2} q_n}.$$

For our applications, continued fractions with positive integer quotients are essential. So we assume now that $a_i > 0$ for $i \geq 1$ (a_0 could be negative) and that a_i are integers, in which case the continued fraction is said to be *simple*. For simple continued fractions we have the following conclusion.

Theorem 6.8.5 *Let the symbols be the same as before and* $[a_0, ..., a_N]$ *a simple continued fraction. Then*

1. *the even convergents* p_{2n}/q_{2n} *increase strictly, while the odd convergents* p_{2n+1}/q_{2n+1} *decrease strictly;*

2. *every odd convergent is greater than any of its corresponding even convergents;*

3. $\frac{p_{2k}}{q_{2k}} \leq [a_0, a_1, ..., a_N] \leq \frac{p_{2k+1}}{q_{2k+1}}.$

Proof: By Theorem 6.8.4 $p_n/q_n - p_{n-2}/q_{n-2}$ has sign $(-1)^n$. This proves part one. Again by Theorem 6.8.4 $p_n/q_n - p_{n-1}/q_{n-1}$ has sign $(-1)^{n-1}$. So $p_{2m+1}/q_{2m+1} > p_{2m}/q_{2m}$. From this inequality and the conclusions of part one the conclusion of part two follows. Finally, p_N/q_N is the greatest of the even, or the least of odd convergents, and part three is true in either case.   □

Theorem 6.8.6 *Let* p_n/q_n *be a convergent to a simple continued fraction. Then* p_n *and* q_n *are relatively prime.*

Proof: By Theorem 6.8.4, $d = 1$ if d divides both p_n and q_n.   □

Obviously, any simple continued fraction $[a_0, a_1, ..., a_N]$ represents a rational number. The converse is the following.

Theorem 6.8.7 *Any rational number can be expressed as a finite simple fraction in just two ways as follows*

$$[a_0, a_1, ..., a_N] = [a_0, a_1, ..., a_N - 1, 1] \quad \text{when } a_N \geq 2$$

or

$$[a_0, a_1, ..., a_{N-1}, 1] = [a_0, a_1, ..., a_{N-1} + 1] \quad \text{when } a_N = 1.$$

Proof: Let U/V be a given rational number. Set $r_{-1} = U, r_0 = V$. By Euclidean algorithm

$$r_{i-2} = a_{i-1}r_{i-1} + r_i, \quad \text{for } i = 1, 2, ..., N+1, \tag{6.59}$$

where $r_{N+1} = 0$, we get a continued fraction $[a_0, a_1, ..., a_N]$, which represents the rational number U/V. In addition, its ith convergent p_i/q_i is determined by (6.57). □

It is not difficult to prove that $|U/V - p_i/q_i|$ decreases monotonically with i, and

$$|U/V - p_i/q_i| \le 1/q_i q_{i+1} < 1/q_{i+1}^2$$

since $q_i > q_{i-1}$ for $i > 1$. The following theorem is true [42, p.262].

Theorem 6.8.8 *If $|p/q - U/V| < 1/2q^2$, then p/q is convergent to U/V.*

Another useful result for deriving the decoding algorithm is the following.

Theorem 6.8.9 *Let the symbols be the same as before.*

1. *If p_i/q_i are convergents to U/V, then $q_{i-1}r_{i+1} + q_i r_i = V$ and $p_{i-1}r_{i+1} + p_i r_i = U$.*

2. *$q_i U - p_i V = (-1)^i r_{i+1}$.*

Now we go back to the decoding problem of the arithmetic residue codes. Recall formula (6.56):

$$x' - x = M_r B'/C.$$

From this formula it follows that

$$\left| \frac{x'}{M_r} - \frac{B'}{C} \right| = \frac{x}{M_r}.$$

If $x/M_r < 1/2C^2$, by Theorem 6.8.8 B'/C is a convergent of the rational number x'/M_r. Then by the Euclidean algorithm of (6.59) and the iterative algorithm described by (6.57) we can find the convergents, one of which must be the error fraction B'/C. By Theorem 6.8.9

$$\frac{U}{V} - \frac{p_i}{q_i} = \frac{(-1)^i r_{i+1}}{q_i V}.$$

Set $U = x' = x + M_r B'/C, V = M_r$, and $p_i/q_i = B'/C$, we have

$$x = (-1)^i r_{i+1}/q_i. \tag{6.60}$$

This is the information number under the assumption that B'/C is the convergent to x'/M_r. For many large x the rational number B'/C is not a convergent to x'/M_r.

Let $C_{max} = \prod_{n+i+1}^{n+2t} m_i$. If $x < M_r/2C_{max}^2$, then $x/M_r < 1/2C^2$, and B'/C must be a convergent to x'/M_r. For the information numbers in the range $x \geq M_r/2C_{max}^2$ the rational number B'/C could not be a convergent to x'/M_r. Let $\delta = 1/C_{max}^2$. If B'/C is not a convergent to x'/M_r, then $x > M_r \delta/2$. This means that x is quite large. In this case we note that

$$\frac{x' - u - (x - u)}{M_r} = \frac{B'}{C}$$

and therefore

$$\left| \frac{x' - u}{M_r} - \frac{B'}{C} \right| = \frac{|x - u|}{M_r}.$$

Thus, if we subtract a proper positive integer u from x, where u is less than x, we can ensure that $(x - u)/M_r < \delta/2$, in which case B'/C is a convergent to $(x' - u)/M_r$. Then we can apply the above iterative procedure to $U = (x' - u)$ and $V = M_r$. This is the idea of the decoding algorithm.

Similarly by Theorem 6.8.9, we can prove the following theorem due to Mandelbaum [54].

Theorem 6.8.10 *If B'/C is a convergent p_i/q_i to $(x' - \lceil M_r \delta/2 \rceil - k \lceil M_r \delta \rceil)/M_r$, then the correct information number is given by*

$$x = (-1)^i r_{i+1}/q_i + \lceil M_r \delta/2 \rceil + k \lceil M_r \delta \rceil.$$

Proof: Let $u = \lceil M_r \delta/2 \rceil + k \lceil M_r \delta \rceil$. By Theorem 6.8.9 we have

$$\frac{x + M_r B'/C - u}{M_r} - \frac{B'}{C} = \frac{(-1)^i r_{i+1}}{q_i M_r}.$$

It follows that

$$x - u = (-1)^i r_{i+1}/q_i.$$

□

Let $x'' = x' - \lceil M_r \delta \rceil$. Theorem 6.8.10 gives a test for determining the correct convergent B'/C using the iterative algorithms (6.57) and (6.59). If

$(-1)^i r_{i+1}/q_i$ is an integer less than $M_r/R - \lceil M_r\delta \rceil$ for $U = x''$ and $V = M_r$, then $p_i/q_i = B'/C$, and we automatically have the correct decoded information number x. Thus, we go through the iterative passes (6.57) and (6.59) with $U = x''$ and $V = M_r$ until we obtain either a legitimate x or a $q_j > C_{max}$. In the latter case we subtract $\lceil M_r\delta \rceil$ from x'' and repeat the iteration, etc. This is the decoding procedure described in [7] which improves the one in [54]. At this stage one can write down the decoding algorithm easily. It is a combination of the CRA and the Euclidean algorithm as well as the iterative algorithm of (6.57). The combination of the iterative algorithms of (6.59) and (6.57) is called the continued fraction algorithm. The first step of this decoding algorithm is to use the CRA to compute x' from the received vector [7].

The above decoding procedure can be improved, as done by Mandelbaum [56]. Note that some of the convergents of x''/M_r, $(x'' - k\lceil M_r\delta \rceil)/M_r$, and therefore B'/C should be the same. This means that we do not start at the beginning of the iterations (6.57) and (6.59) each time we increase k by one. By Theorem 6.8.7, B'/C should have the same convergents except, possibly, the last two as the x''/M_r, $(x'' - k\lceil M_r\delta \rceil)/M_r$ that leads to the correct convergent $p/q = B'/C$. Also, the initial s convergents (for some $s > 0$) of x''/M_r should be the same as the initial s convergents of B'/C. These are the key observations of Mandelbaum which lead to an improvement of the above decoding procedure. For further details about the improved algorithm by Mandelbaum we refer to [56].

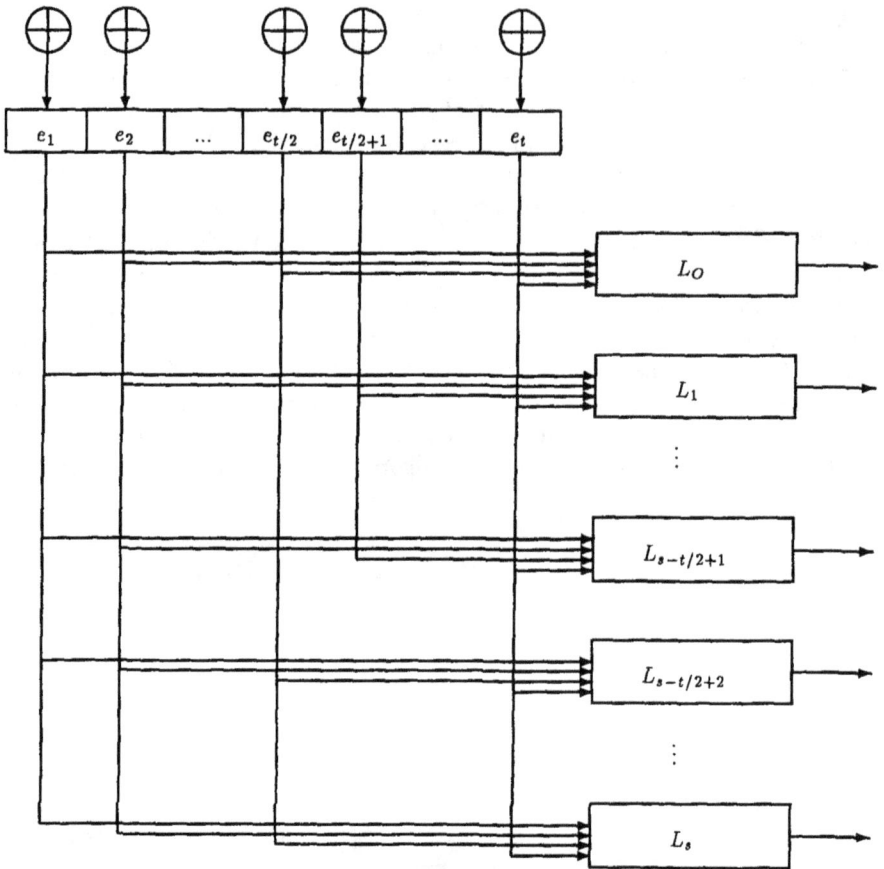

Figure 6.1: The Bossen-Yau decoder.

Chapter 7

In Cryptography

Secret sharing is an important topic in cryptography. One of the most straight-forward applications of the CRT is in secret sharing, as it constitutes a secret sharing scheme without any modification. Error-correcting codes based on the CRT can be directly used as secret-sharing schemes, and conversely, secret-sharing threshold schemes may also be used as error-correcting codes. Bridging some codes and secret-sharing threshold schemes is another topic of this chapter. We shall see that MDS error-correcting codes and linear secret-sharing schemes are equivalent.

The CRT can also be used to construct keystream generators over residue class rings, as shown in this chapter. In this chapter we investigate also relations between some knapsack problems and the CRA, and present some easy knapsack problems. Two public-key cryptosystems based on the CRT are also discussed.

7.1 Secret Sharing and CRT

In this section we are concerned with a number of applications of the CRT and CRA in secret sharing. The idea of applying the CRT in secret sharing is similar to that in coding.

The CRT is itself a secret sharing scheme without any modification. Let $m_1, m_2, ..., m_t$ be t pairwise relatively prime positive integers, and $m = \prod_{i=1}^{t} m_i$. Consider now the CRT for integers with respect to these moduli. Suppose that we have a secret which is an integer s with $0 \leq s < m$. The secret s can be shared among t parties as follows. Let $P_1, P_2, ..., P_t$ denote the t parties who are going to share the secret. We give P_i the residue $s_i = s \bmod m_i$ as the secret information which is only known to P_i. By the CRT the t pieces of information s_i are sufficient to determine the original secret s, but any set of less than t residues s_i cannot determine the original s. This is an secret-sharing scheme, but not a threshold scheme.

157

A (k,t) *secret-sharing threshold scheme* is defined as follows. t parties P_i share a secret s with the following properties:

1. Each party has a share s_i about the secret s which is not known to other parties.

2. The secret s can be "easily" computed from any k shares s_i.

3. No $k-1$ shares s_i give any information about the secret s.

In terms of information terminology, the above second condition means that the mutual information between s and any k residues s_i is equal to the self information of s, i.e., $I(s; (s_{i_1}, ..., s_{i_k})) = I(s)$, where $1 \le i_1 < i_2 < \cdots < i_k \le t$. A customary definition for the requirement in the second condition is the existence of a polynomial algorithm for the calculation of s from any k shares s_i. The above third condition says that the mutual information between s and any $k-1$ shares s_i is zero, i.e., $I(s; (s_{i_1}, ..., s_{i_{k-1}})) = 0$, where $1 \le i_1 < i_2 < \cdots < i_{k-1} \le t$. In what follows we shall describe several secret-sharing schemes based on the CRT and CRA.

Scheme I:

Let m_i, $i = 1, ..., t$, be t pairwise relatively prime integers no less than 2. Let

$$\min(k) = \min_{1 \le i_1 < i_2 < \cdots < i_k \le t} m_{i_1} m_{i_2} \cdots m_{i_k},$$
$$\max(k-1) = \max_{1 \le i_1 < i_2 < \cdots < i_{k-1} \le t} m_{i_1} m_{i_2} \cdots m_{i_{k-1}},$$

where $1 < k \le t$. Choose w to be the largest positive integer such that

- $w < \frac{\min(k)}{\max(k-1)}$; and

- $\gcd(w, m_i) = 1$, $i = 1, 2, \cdots, t$.

Define

$$m = \min(k).$$

In the secret-sharing scheme the secret is an integer s with $0 \le s < w$. Under the assumption that the secret is equally likely to be any integer between 0 and $w - 1$, the self information or uncertainty of the secret is $\log_2 w$.

The shares for the t parties are calculated as follows. Choose an integer a such that $0 \le s + aw < m$ and let

$$s' = s + aw.$$

The shares are then given by

$$s_i = s' \bmod m_i, \quad i = 1, 2, \cdots, t,$$

where s_i is the share of party P_i.

Now we prove that any $k - 1$ or fewer shares give no information about the secret, but any k or more shares determine the secret. Without loss of generality, suppose that $s_1, s_2, ..., s_h$ are known, where $1 \leq h \leq t$. Let $M = \prod_{i=1}^{h} m_i$ and $M_j = M/m_j$ for $j = 1, ..., h$. Then M_j and m_j are relatively prime. With Euclidean algorithm we can find two integers u_j and v_j such that

$$M_j u_j + m_j v_j = 1.$$

By the CRA we compute an integer

$$s'' = \sum_{j=1}^{h} M_j u_j s_j \bmod M,$$

where $0 \leq s'' < M$.

If $h \geq k$, then $M \geq \min(k) = m > w$. Thus by the CRT $s'' = s'$ and the secret is given by $s = s'' \bmod w$.

If $h = k - 1$, then $M \leq \max(k-1) < m/w$. In this case s' must be expressed of the form

$$s' = s'' + bM,$$

where b is an integer such that $0 \leq s' < m$. It follows that

$$-1 < -\frac{s''}{M} \leq b < \frac{m - s''}{M}.$$

Note that

$$\frac{m - s''}{M} > \frac{m - M}{M} = \frac{m}{M} - 1 > w - 1.$$

The uncertainty of s' is no less than $\log_2 w$ bits. Since w and M are relatively prime, we have (see Section 3.6)

$$I(s; (s_1, s_2, \cdots, s_{k-1})) = 0.$$

Thus, any $k - 1$ or fewer shares give no information about the secret s. Summarizing the above results proves that this is a (k, t) threshold scheme. □

To have a better understanding of the above threshold system, we consider an example. Choose $k = 3, t = 4, m_1 = 5, m_2 = 7, m_3 = 11, m_4 = 13$. With the above notations we have

$$\min(k) = m = 385, \quad \max(k - 1) = 143, \quad w = 2.$$

In this example the secret is either 0 or 1. Thus the uncertainty or self information of the secret is one bit. To compute the four shares, choose randomly an integer a with $0 \le a \le 192$ and let

$$s' = s + 2a.$$

The four shares are given by $s_i = s' \bmod m_i$.

Suppose that we are given two shares $s_1 = 1$ and $s_2 = 5$. We now see how much information the two shares give about the secret. As before, let $M = m_1 m_2 = 35$, $M_1 = 7$ and $M_2 = 5$. Then with Euclidean algorithm we get

$$7 \times (-2) + 5 \times 3 = 1.$$

By the CRA we have

$$s'' = 7 \times (-2) \times s_1 + 5 \times 3 \times s_2 \bmod 35 = 26.$$

Hence

$$s' = 26 + 35b,$$

where $0 \le b \le 10$. It follows that

$$s = s' \bmod 2 = b \bmod 2.$$

Since b ranges from 0 to 11, s takes on 1 and 0 equally likely. Thus, the two shares give no information about the secret.

Scheme II:

Threshold secret-sharing schemes based on the CRT for polynomials are more attractive. Let $F = GF(q)$ be a finite field, and $m_i(x)$, $i = 1, ..., t$, be t pairwise relatively prime polynomials of $F[x]$ with degree no less than 1. Define

$$\min[k] = \min_{1 \le i_1 < i_2 < \cdots < i_k \le t} \deg(m_{i_1} m_{i_2} \cdots m_{i_k}),$$

$$\max[k - 1] = \max_{1 \le i_1 < i_2 < \cdots < i_{k-1} \le t} \deg(m_{i_1} m_{i_2} \cdots m_{i_{k-1}}),$$

where $1 < k \le t$. Let w be the largest positive integer such that

- there is a polynomial $W(x)$ of degree w over F with $\gcd(W(x), m_i(x)) = 1$ for $i = 1, 2, \cdots, t$; and

- $w \leq \frac{\min[k]}{\max[k-1]}$.

Choose such a polynomial $W(x)$ over F.

In the secret-sharing scheme the secret is a polynomial $s(x) \in F[x]$ of degree less than w. Thus, the self information or uncertainty of the secret is $w \log_2 q$ bits. The shares for the t parties are computed as follows. Choose any polynomial $a(x) \in F[x]$ such that

$$\deg(s(x) + a(x)W(x)) < \min[k]$$

and let

$$s'(x) = s(x) + a(x)W(x).$$

The shares are then given by

$$s_i(x) = s'(x) \bmod m_i(x), \quad i = 1, 2, \cdots, t.$$

Similarly, we can prove that this is a (k, t) threshold scheme. Suppose that k shares $s_1(x), s_2(x), \cdots, s_k(x)$ are given, the computation of the secret is similarly performed as follows.

Let $M(x) = \prod_{i=1}^{k} m_i(x)$ and $M_j(x) = M(x)/m_j(x)$ for $j = 1, ..., k$. Then $M_j(x)$ and $m_j(x)$ are relatively prime. With Euclidean algorithm we can find two polynomials $u_j(x)$ and $v_j(x)$ from $F[x]$ such that

$$M_j(x)u_j(x) + m_j(x)v_j(x) = 1.$$

With the CRA we compute

$$s'(x) = \sum_{j=1}^{k} M_j(x)u_j(x)s_j(x) \bmod M(x).$$

The secret is then given by

$$s(x) = s'(x) \bmod W(x).$$

The algorithm for the calculation of the secret is efficient.

A special case of the above secret-sharing scheme is when the moduli $m_i(x) = x - a_i$, where $a_1, ..., a_t$ are t distinct elements of the field F. This special case is the secret-sharing scheme based on polynomial interpolation suggested by Shamir [81], since $f(x) \bmod (x - a) = f(a)$, where $f(x) \in F[x]$ and $a \in F$. By choosing different sets of special moduli we get different threshold schemes.

Scheme III:

The generalized CRA for integers described by Theorem 2.5.1 in Section 2.5 can also be applied to design secret-sharing threshold schemes. Choose integers $m_1, ..., m_t$ and $a_1, ..., a_t$ such that

$$\gcd(m_i, m_j) = 1, \quad i \neq j,$$
$$\gcd(a_i, m_i) = 1, \quad i = 1, 2, ..., t.$$

Let

$$\min(k) = \min_{1 \leq i_1 < i_2 < \cdots < i_k \leq t} m_{i_1} m_{i_2} \cdots m_{i_k},$$
$$\max(k-1) = \max_{1 \leq i_1 < i_2 < \cdots < i_{k-1} \leq t} m_{i_1} m_{i_2} \cdots m_{i_{k-1}},$$

where $1 < k \leq t$. Choose w to be the largest positive integer such that

- $w < \frac{\min(k)}{\max(k-1)}$; and

- $\gcd(w, m_i) = 1, \quad i = 1, 2, \cdots, t.$

Define

$$m = \min(k).$$

In the secret-sharing scheme the secret is an integer s with $0 \leq s < w$. The shares for the t parties are computed as follows. Choose an integer a such that $0 \leq s + aw < m$, and let $s' = s + aw$. The shares are then given by

$$s_i = a_i s' \bmod m_i, \quad i = 1, 2, \cdots, t,$$

where s_i is the share of party P_i.

By Theorem 2.5.1 there is exactly one solution modulo $m_{i_1} \cdots m_{i_l}$ to the following system of congruences

$$a_{i_j} x = s_{i_j} \bmod m_{i_j}, \quad j = 1, 2, ..., l$$

for any set of distinct integers $1 \leq i_1, i_2, ..., i_l \leq t$, where $k \leq l \leq t$. Since a_i and m_i are relatively prime, by Euclidean algorithm we can find integers c_i such that $a_i c_i = 1 \bmod m_i$. Suppose that $s_1, s_2, ..., s_k$ are known, let s' be the unique solution of the system of congruences

$$x = c_i s_i \bmod m_i, \quad i = 1, 2, ..., k.$$

This is solved by the CRA for integers. The secret is then given by

$$s = s' \bmod w.$$

It is easily seen that this is a (k, t) secret-sharing threshold scheme, which is a generalization of Scheme I and more flexible. The flexibility of the system is due to the fact that the parameters $a_1, a_2, ..., a_t$ can be controlled by parties other than those t parties who are assumed to share the secret.

Scheme IV:

Similarly, the generalized CRA for polynomials described by Theorem 2.5.3 in Section 2.5 can also be used for secret sharing as follows. Choose polynomials $m_1(x), ..., m_t(x)$ and $a_1(x), ..., a_t(x) \in F[x]$ such that

$$\gcd(m_i(x), m_j(x)) = 1, \quad i \neq j,$$
$$\gcd(a_i(x), m_i(x)) = 1, \quad i = 1, 2, ..., t.$$

Let

$$\min[k] = \min_{1 \leq i_1 < i_2 < \cdots < i_k \leq t} \deg(m_{i_1} m_{i_2} \cdots m_{i_k}),$$
$$\max[k-1] = \max_{1 \leq i_1 < i_2 < \cdots < i_{k-1} \leq t} \deg(m_{i_1} m_{i_2} \cdots m_{i_{k-1}}),$$

where $1 < k \leq t$. Let w be the largest positive integer such that

- there is a polynomial $W(x)$ of degree w over F with $\gcd(W(x), m_i(x)) = 1$ for $i = 1, 2, \cdots, t$; and

- $w \leq \frac{\min[k]}{\max[k-1]}$.

Choose such a polynomial $W(x)$ over F.

In this secret-sharing scheme the secret is a polynomial $s(x) \in F[x]$ with $0 \leq \deg(s(x)) < w$. The shares are computed as follows. Choose a polynomial $a(x) \in F[x]$ such that

$$\deg(s(x) + a(x)W(x)) < \min[k]$$

and define

$$s'(x) = s(x) + a(x)W(x).$$

The shares are then given by $s_i(x) = a_i(x)s'(x) \bmod m_i(x)$ for each i.

By Theorem 2.5.3 there is exactly one solution modulo $m_{i_1} \cdots m_{i_l}$ to the following system of congruences

$$a_{i_j}(x)u(x) = s_{i_j}(x) \bmod m_{i_j}(x), \quad j = 1, 2, ..., l$$

for any set of distinct integers $1 \leq i_1, i_2, ..., i_l \leq t$, where $k \leq l \leq t$. Since $a_i(x)$ and $m_i(x)$ are relatively prime, by Euclidean algorithm we can find polynomials $c_i(x)$ such that $a_i(x)c_i(x) = 1 \bmod m_i(x)$. Suppose that $s_1(x), s_2(x), ..., s_k(x)$ are known, let $s'(x)$ be the unique solution of the system of congruences

$$u(x) = c_i(x)s_i(x) \bmod m_i(x), \quad i = 1, 2, ..., k.$$

This is solved by the CRA for polynomials. The secret is then given by $s(x) = s'(x) \bmod W(x)$. It is easily seen that this is a (k, t) secret-sharing threshold scheme, which is a generalization of Scheme II and more flexible.

Scheme V:

We can construct another secret-sharing threshold scheme based on the polynomial interpolation over rings $\mathbf{Z}/(m)$. Recall now the definitions and symbols of Section 4.3. To apply the polynomial interpolation algorithm described in Section 4.3 to secret-sharing, we need the following result.

Theorem 7.1.1 *Let the notations be the same as in Section 4.3. Assume that $n_j = p_j$ for $j = 1, 2, ..., t$. If there is a polynomial $U(x) \in \mathbf{Z}/(m)[x]$ such that $U(\alpha_i) = \beta_i$ for $i = 0, 1, ..., n - 1$, then there is exactly one such polynomial of degree no more than $p_t - 1$. Here the uniqueness is understood in the sense of both polynomial functions and formal polynomials.*

Proof: Since $n_i = p_i$, by the proof of Theorem 4.3.1 we can construct such a polynomial of degree $p_t - 1$ if there exists one. The uniqueness in the two senses follows from the facts that $n_i = p_i$ for all i and that $\mathbf{Z}/(p_i)$ are fields, as well as the CRT.   □

One application of the polynomial interpolation over $\mathbf{Z}/(m)$ is the following secret-sharing scheme which is an extension of the Shamir secret-sharing scheme based on polynomial interpolation over $\mathbf{Z}/(p)$, where p is a prime [81]. Let m and p_i be the same as before. Choose n elements $\alpha_1, \alpha_2, ..., \alpha_n \in \mathbf{Z}/(m)$ such that

1. for any k of them, say $\alpha_{i_1}, \alpha_{i_2}, ..., \alpha_{i_k}$, the following holds

$$\{\alpha_{i_s} \bmod p_j : s = 1, 2, ..., k\} = \mathbf{Z}/(p_j), j = 1, 2, ..., t;$$

2. for any $k - 1$ elements α_i, say $\alpha_{i_1}, \alpha_{i_2}, ..., \alpha_{i_{k-1}}$, there is at least one j such that

$$\{\alpha_{i_s} \bmod p_j : s = 1, 2, ..., k - 1\} \neq \mathbf{Z}/(p_j).$$

The secret to be shared among n parties is an integer s_0 with $0 \leq s_0 < m$. To compute the shares choose randomly and independently $p_t - 1$ integers $s_1, ..., s_{p_t-1}$, where $0 \leq s_i < m$. Let $S(x) = \prod_{i=0}^{p_t-1} s_i x^i$. The share of the ith party is the value $\beta_i = S(\alpha_i)$. It is not hard to see that any $k - 1$ β_i give no information about s_0, but with any k of them the interpolation algorithm described in Section 4.3 recovers $S(x)$ and therefore the secret s_0. Thus, this is a (k, n) threshold scheme.

We now consider one example. Choose $p_1 = 3, p_2 = 5$ and $m = p_1 p_2 = 15$. Let $n = 6, k = 6$ and $\alpha_i = i - 1$ for $i = 1, 2, ..., 6$. Then the above two conditions are satisfied. The secret to be shared is an integer s_0 with $0 \leq s_0 < 15$. Choose s_1, s_2, s_3, s_4 from $\mathbf{Z}/(15)$ randomly and independently. Let

$$S(x) = \sum_{i=0}^{4} s_i x^i.$$

The secret information about s_0 given to the ith party is

$$\beta_i = S(\alpha_i)$$

This is a $(6,6)$ threshold scheme.

It may not be so easy to find a set of elements $\alpha_1, \alpha_2, ..., \alpha_n \in Z/(m)$ such that the above two conditions hold. But it is easy to satisfy Condition 1. If Condition 2 is not satisfied, this is not a threshold scheme, but certainly a secret-sharing scheme.

In general, the CRT in any Euclidean domain can be applied to construct secret-sharing schemes. Schemes I and II are suggested by Asmuth and Bloom [5].

7.2 Secret Sharing and Codes

Some error-correcting codes and secret-sharing schemes are closely related. This is demonstrated by the redundant residue codes described in Chapter 6 and the secret-sharing schemes of the preceding section. One common feature is that they are all based on the CRT and CRA. In this section we shall have a look at the relations between some error-correcting codes and secret-sharing schemes.

Recall first the Reed-Solomon codes described in Section 6.3. Let α be a generating element of the finite field $GF(q)$, and $n = q - 1$. The Reed-Solomon code is described by

$$C = \{(u(1), u(\alpha), \cdots, u(\alpha^{n-1}))|u(x) \in GF(q)[x]_k\},$$

where $GF(q)[x]_k$ consists of all polynomials of degree less than k of $GF(q)[x]$. As we saw, the Reed-Solomon code is a special redundant residue code with the set of special moduli $m_i(x) = x - \alpha^i$. It is an $[n, k, n-k+1]$ linear MDS code, which can be easily used to construct a (k, n) threshold scheme based on polynomial interpolation suggested by Shamir [81]. The secret to be shared is an element $u_0 \in GF(q)$. To compute the shares, choose randomly and independently $k-1$ elements $u_1, u_2, \cdots, u_{k-1} \in GF(q)$. Let $u(x) = \sum_{i=0}^{k-1} u_i x^i$, the share of party P_i is given by

$$s_i = u(\alpha_i), \quad i = 1, 2, ..., n.$$

Since the dual code of the Reed-Solomon code is an $[n, n - k, k + 1]$ BCH MDS code by Theorem 6.3.3, it is not hard to see that this is an $(n - k, n)$ secret-sharing threshold scheme. The CRA for polynomials can be easily used to recover the secret if k or more s_i are known. Thus, computing the shares s_i is encoding, and the secret recovering can be done by solving simultaneous linear equations or by the CRA for polynomials, which are easier than decoding Reed-Solomon codes, since in the context of secret-sharing we know that in certain positions there are no errors at all.

One can also use the extended $[n + 1, k, n - k + 2]$ Reed-Solomon code described in Section 6.3 as a $(k + 1, n + 1)$ secret-sharing threshold scheme. The generalized Reed-Solomon code over $GF(q^m)$, denoted by $GRS_k(\Delta, \mathbf{b})$ in Section 6.6, can also be used to construct threshold schemes, belonging to Scheme IV of the last section. The Bossen-Yau codes can also be used as secret-sharing schemes, belonging to Scheme II. Other redundant residue codes can be similarly employed as secret-sharing schemes.

The arithmetic codes described in Section 6.8 are similarly employed as secret-sharing threshold schemes, belonging to Scheme I of the last section, and their generalized codes are secret-sharing threshold schemes corresponding to Scheme III. Conversely, those secret-sharing schemes are also error-correcting codes.

There is another approach to the construction of secret-sharing schemes based on linear error-correcting codes, which will be referred to as the second approach in the sequel. Let $\mathcal{C}$ be an $[n + 1, k, d]$ code over $GF(q)$, and $G = [\mathbf{g}_0 \mathbf{g}_1 \cdots \mathbf{g}_n]$ be a generator matrix of $\mathcal{C}$. Furthermore, let $\mathbf{g}_0 = (g_{10} g_{20} \cdots g_{k0})^T \neq \mathbf{0}$ be the first column of G. Suppose that the secret is an element of $GF(q)$.

In the secret-sharing scheme based on $\mathcal{C}$ with respect to the second approach we choose randomly a vector $\mathbf{s} = (s_1, s_2, \cdots, s_k) \in GF(q)^k$ such that

$$s = \mathbf{s} \mathbf{g}_0 = \sum_{i=1}^{k} s_i g_{i0}.$$

It is easy to see there are q^{k-1} choices for $\mathbf{s}$. To compute the shares, we take $\mathbf{s} = (s_1, s_2, \cdots, s_k)$ as the information vector, and calculate the codeword corresponding to $\mathbf{s}$ as follows:

$$\mathbf{t} = (t_0, t_1, \cdots, t_n) = \mathbf{s} G. \tag{7.1}$$

By the choice of $\mathbf{s}$, $t_0 = s$ is the secret. The other coordinates $t_1, t_2, \cdots, t_n$ are shares and t_i is given only to party P_i for each $i \geq 1$.

To prove that this is a threshold scheme, we need the following result which is equivalent to the characterization given in [58].

Theorem 7.2.1 *Let the symbols be the same as before and assume that the first column $\mathbf{g}_0$ of the generator matrix is a linear combination of the other columns $\mathbf{g}_1, \cdots, \mathbf{g}_n$. Then a set of shares $\{t_{i_1}, t_{i_2}, \cdots, t_{i_m}\}$ can determine the secret $t_0 = s$ if and only if $\mathbf{g}_0$ is a linear combination of $\mathbf{g}_{i_1}, \mathbf{g}_{i_2}, \cdots, \mathbf{g}_{i_m}$, where $1 \le i_1 < i_2 < \cdots < i_m \le n$ and $m \le n$.*

Proof: Assume that $\mathbf{g}_0$ is a linear combination of these $\mathbf{g}_{i_j}$, let

$$\mathbf{g}_0 = \sum_{j=1}^{m} x_j \mathbf{g}_{i_j}. \tag{7.2}$$

we have then

$$\begin{aligned} t_0 &= s\mathbf{g}_0 = \sum_{j=1}^{m} x_j s\mathbf{g}_{i_j} \\ &= \sum_{j=1}^{m} x_j t_{i_j}. \end{aligned} \tag{7.3}$$

Thus, the secret $s = t_0$ is recovered, given the shares $t_{i_1}, \cdots, t_{i_m}$.

Assume now that $\mathbf{g}_0$ is not a linear combination of the vectors $\mathbf{g}_{i_1}, \cdots, \mathbf{g}_{i_m}$. By the assumption that $\mathbf{g}_0$ is a linear combination of $\mathbf{g}_1, \mathbf{g}_2, \cdots, \mathbf{g}_n$ the set of column vectors $\{\mathbf{g}_{i_1}, \cdots, \mathbf{g}_{i_m}\}$ must be extended into another set $\{\mathbf{g}_{i_1}, \cdots, \mathbf{g}_{i_m}, \mathbf{g}_{i_{m+1}}, \cdots, \mathbf{g}_{i_{m+t}}\}$ such that $\mathbf{g}_0$ is a linear combination of these vectors, where $m < m + t \le n$. Let

$$\mathbf{g}_0 = \sum_{j=1}^{m+t} x_j \mathbf{g}_{i_j}. \tag{7.4}$$

At least one of the constants $x_{m+1}, x_{m+2}, \cdots, x_{m+t}$ must be nonzero since $\mathbf{g}_0$ is not a linear combination of the vectors $\mathbf{g}_{i_1}, \cdots, \mathbf{g}_{i_m}$. It follows from (7.4) that

$$\begin{aligned} t_0 &= s\mathbf{g}_0' = \sum_{j=1}^{m+t} x_j s\mathbf{g}_{i_j} \\ &= \sum_{j=1}^{m+t} x_j t_{i_j}. \end{aligned} \tag{7.5}$$

Since at least one of the constants $x_{m+1}, x_{m+2}, \cdots, x_{m+t}$ is nonzero and $t_{i_{m+1}}, \cdots, t_{i_{m+t}}$ are unknown, by (7.5) t_0 is equally likely any element of $GF(q)$, and thus the set of shares $\{t_{i_1}, \cdots, t_{i_m}\}$ gives no information about the secret. $\square$

The proof of Theorem 7.2.1 also shows how to compute the secret when a set of shares is given. It also proves that in such a secret-sharing scheme a set of shares either determines the secret or gives no information about the secret. Such a secret-sharing scheme is said to be *perfect* .

Recall that for any $[n, k, d]$ linear code over a field, $d \le n - k + 1$. Codes with $d = n - k + 1$ are called MDS (maximum distance separable), since such a code has the maximum possible distance between codewords, and the codewords can be separated into message symbols and check symbols. A number of redundant codes belong to this class. One important aspect of this class of linear codes is in secret-sharing, as shown by the following result.

Theorem 7.2.2 *The secret-sharing scheme based on any $[n, k, n-k+1]$ MDS code over $GF(q)$ with respect to the second approach is a $(k, n-1)$ threshold scheme.*

To prove this conclusion, we need to prove a number of properties of MDS codes which are summarized in the following two lemmas [51, pp.318-319].

Lemma 7.2.3 *Let C be an $[n, k, d]$ code over $GF(q)$ with parity matrix H and generator matrix G.*

1. *C is MDS if and only every $n-k$ columns of H are linearly independent.*

2. *If C is MDS, so is the dual code $C^{\perp}$.*

Proof: By definition it is straightforward to see that C contains a codeword of weight w if and only if w columns of H are linearly dependent. Therefore C has $d = n - k + 1$ if and only if no $n - k$ or fewer columns of H are linearly dependent. This proves part one.

By definition H is a generator matrix of $C^{\perp}$. By part one any $n - k$ columns of H are linearly independent, so only the zero codeword is zero on as many as $n - k$ coordinates. Therefore $C^{\perp}$ has minimum distance at least $k + 1$, i.e., it has parameters $[n, n-k, k+1]$.   □

Combining the two parts of the above lemma proves the following result.

Lemma 7.2.4 *Let C be an $[n, k, d]$ code over $GF(q)$. The following statements are equivalent:*

1. *C is MDS,*

2. *every k columns of a generator matrix G are linearly independent (i.e., any k symbols of the codewords may be taken as message symbols);*

3. *every $n - k$ columns of a parity check matrix H are linearly independent.*

Proof of Theorem 7.2.2: Part two of Lemma 7.2.4 says that every k columns of G are linearly independent. By Theorem 7.2.1 any $k - 1$ shares give no information about the secret. Since the generator matrix G has rank k and every k columns of G are linearly independent, $\mathbf{g}_0$ must be a linear combination of any set of k columns of G. By Theorem 7.2.1 any set of k shares determines the secret, and thus this is a $(k, n-1)$ threshold scheme.   □

Theorem 7.2.5 *If the secret-sharing scheme based on an $[n, k, d]$ code C over $GF(q)$ with respect to the second approach is a threshold scheme, then C must be a MDS code, i.e., $d = n - k + 1$.*

Proof: It is not hard to see that it is a $(k, n-1)$ threshold scheme. Thus it suffices to prove that every k columns of $G = [\mathbf{g}_0 \mathbf{g}_1 \cdots \mathbf{g}_{n-1}]$ are linearly independent.

We first prove that for every set of indices $1 \le i_1 < i_2 < \cdots i_k \le n-1$, the columns $\mathbf{g}_{i_1}, \mathbf{g}_{i_2}, \cdots, \mathbf{g}_{i_k}$ must be linearly independent. Suppose they are linearly dependent, without loss of generality let

$$\mathbf{g}_{i_1} = \sum_{j=2}^{k} y_j \mathbf{g}_{i_j}. \tag{7.6}$$

Since the shares $t_{i_1}, \cdots, t_{i_k}$ can determine the secret, by Theorem 7.2.1 $\mathbf{g}_0$ is a linear combination of $\mathbf{g}_{i_1}, \mathbf{g}_{i_2}, \cdots, \mathbf{g}_{i_k}$. Let

$$\mathbf{g}_0 = \sum_{j=1}^{k} z_j \mathbf{g}_{i_j}, \tag{7.7}$$

where all the coefficients z_j are nonzero, otherwise some $k-1$ shares can determine the secret by Theorem 7.2.1. Equations (7.6) and (7.7) together show that $\mathbf{g}_0$ is a linear combination of $\mathbf{g}_{i_2}, \mathbf{g}_{i_3}, \cdots, \mathbf{g}_{i_k}$. Again by Theorem 7.2.1 the $k-1$ shares $t_{i_2}, t_{i_3}, \cdots, t_{i_k}$ can determine the secret. This is contrary to the definition of $(k, n-1)$ threshold schemes. Thus $\mathbf{g}_{i_1}, \mathbf{g}_{i_2}, \cdots, \mathbf{g}_{i_k}$ must be linearly independent.

It can be similarly proven that any set of k columns of G containing $\mathbf{g}_0$ must be linearly independent. This completes the proof.   □

To have more secret-sharing threshold schemes, we need more MDS codes. As we saw in Chapter 6, Reed-Solomon codes, extended Reed-Solomon codes, and generalized Reed-Solomon codes are MDS. Finding more MDS codes is equivalent to solving the following problem:

Problem: Given k and n, where $n > k$, find $k \times n$ matrices over $GF(q)$ having every k columns linearly independent.

MDS codes, and therefore secret-sharing threshold schemes, are related to orthogonal arrays. An $m \times n$ matrix A with entries from a set of q elements is called an *orthogonal array* of size m, n constraints, q level, strength k, and index λ if any set of k columns of A contains all q^k possible row vectors exactly λ times. Such an array is denoted by (m, n, q, k). Clearly $m = \lambda q^k$. An example is the following $(4, 3, 2, 2)$ orthogonal array:

$$A = \begin{bmatrix} +1 & +1 & +1 \\ -1 & +1 & -1 \\ -1 & -1 & +1 \\ +1 & -1 & -1 \end{bmatrix}.$$

Binary error-correcting codes, and thus secret-sharing threshold schemes, are intimately related to correlation-immune functions which are used for stream ciphering [85, 30]. Let $x_1, x_2, ..., x_n$ be independent random binary variables which take on 0 and 1 equally likely, and $f(x_1, x_2, ..., x_n)$ be a Boolean function from $GF(2)^n$ to $GF(2)$. Then $z = f(x_1, x_2, ..., x_n)$ is a binary random variable taking on 0 and 1 equally likely. The Boolean function f is said to be mth order *correlation-immune* if for any set of indices $1 \leq i_1 < i_2 < \cdots < i_m \leq n$, the mutual information $I(z; (x_{i_1}, x_{i_2}, ..., x_{i_m})) = 0$, and for at least one set of indices $1 \leq i_1 < i_2 < \cdots < i_{m+1} \leq n$, the mutual information $I(z; (x_{i_1}, x_{i_2}, ..., x_{i_{m+1}})) > 0$, where $I(A; B)$ denotes the amount of mutual information between events A and B, and $I(A)$ the amount of self-information of A. In other words, if f is mth order correlation-immune, then any set of m variables x_i cannot determine the value of $f(x_1, x_2, ..., x_n)$, but there is at least one set of $m+1$ variables which gives some information about the value of $f(x_1, x_2, ..., x_n)$. By definition any set of $t \leq m$ variables cannot determine the value of $f(x_1, x_2, ..., x_n)$. Correlation-immunity is just another way of looking at *essential variables* of functions in the algebra of logic [75].

The function $f(x_1, x_2, ..., x_n) = x_1 + x_2 + \cdots + x_n$ is $(n-1)$th order correlation-immune since any $n - 1$ variables leave the value of f undetermined.

Let C be an $[n, k]$ binary code, and $C^\perp$ its dual code. The *characteristic function* of $C^\perp$ is defined by

$$f_{C^\perp}(x_1, x_2, ..., x_n) = \begin{cases} 1, & x \in C^\perp; \\ 0, & \text{otherwise.} \end{cases}$$

Let the generator matrix of C be

$$G = \begin{bmatrix} g_{11} & g_{12} & \cdots & g_{1n} \\ g_{21} & g_{22} & \cdots & g_{2n} \\ \vdots & \vdots & \vdots & \vdots \\ g_{k1} & g_{k2} & \cdots & g_{kn} \end{bmatrix}.$$

The characteristic function $f_{C^\perp}(x_1, x_2, ..., x_n)$ is then given by

$$f_{C^\perp}(x_1, x_2, ..., x_n) = \prod_{i=1}^{k} \left(\sum_{j=1}^{n} g_{ij} x_j + 1 \right),$$

where the multiplication is the bit multiplication, and addition is the exclusive-or operation. The following result is straightforward.

Theorem 7.2.6 *Let C be an $[n, k]$ binary code, $C^\perp$ the dual code of C. Then the minimum distance of C is d if and only if the characteristic function $f_{C^\perp}$ is $(d - 1)$th order correlation-immune.*

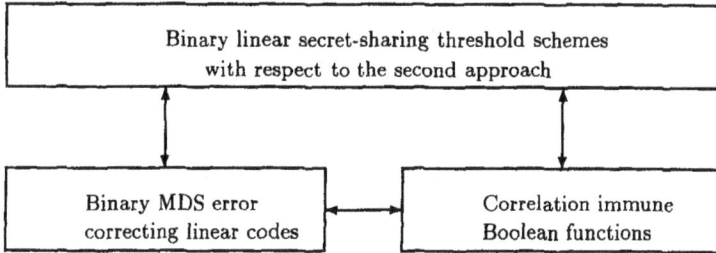

Figure 7.1: The relations between secret-sharing, codes, and Boolean functions.

Combining Theorems 7.2.6 and 7.2.5 proves the following conclusion.

Theorem 7.2.7 *Let C be an $[n, k]$ linear code, and $C^\perp$ the dual code of C. Then the secret-sharing scheme based on C with respect to the second approach is a threshold scheme if and only if the characteristic function $f_{C^\perp}$ is $(n-k)$th order correlation-immune.*

Secret-sharing schemes based on MDS codes can also be studied with a spectral approach since a spectral approach to the correlation immunity of Boolean functions has been carried out by Xiao and Massey [104, 30].

Summarizing the above, we see that secret-sharing threshold schemes based on linear error-correcting codes with respect to the second approach, MDS codes, and correlation-immune functions for stream ciphering are equivalent. Figure 7.1 depicts the one-to-one correspondence.

Threshold schemes are only one kind of secret-sharing schemes, where each party has the same role in sharing the secret. However, in some practical situations some parties should have more information about the secret than others. For more about secret-sharing schemes based on codes we refer to [3, 58].

7.3   CRT and Stream Ciphering

Pseudorandom sequences have wide applications in simulation, sampling, computer programming, ranging systems, global positioning systems, radar systems, spread-spectrum communications systems, and especially in stream ciphers [30]. One random aspect of sequences is the linear complexity (also known as linear span and linear equivalence), which is of great importance in stream ciphers and multiple-access communications systems.

Let R be a ring with multiplicative identity 1, and $s^N = s_0 s_1 \cdots s_{N-1}$ be a sequence of length N over R, where $s_i \in R$. If N is finite, the sequence is called finite; if $N = +\infty$, it is called semi-infinite, and denoted by s^∞. A sequence s^∞ is called periodic if there is an integer N such that $s_{N+j} = s_j$ for each $j \geq 0$, and ultimately periodic if there are two integers N and j_0 such that $s_{N+j} = s_j$ for all $j \geq j_0$.

If s^N satisfies a linear recurrence relation

$$s_i = a_1 s_{i-1} + a_2 s_{i-2} + \cdots + a_l s_{i-l}, \quad i \geq l, \; a_i \in R,$$

then there exits such a shortest linear recurrence relation, and the shortest l is called the linear complexity or linear span of the sequence and is denoted by $L(s^N)$. The linear complexity of a finite sequence s^N is defined to be N if s^N does not satisfy such a linear recurrence relation. For semi-infinite sequences the linear complexity is defined to be $+\infty$ if they satisfy no finite linear recurrence relation. For ultimately periodic sequences the linear complexity is finite. If the linear complexity of a sequence over a field is l, then $2l$ successive characters of the sequence can be used to determine a linear recurrence relation of length l satisfied by the sequence with the Berlekamp-Massey algorithm [57], which has complexity $O(l^2)$. Consequently, $2l$ successive characters of the sequence are sufficient to determine the whole sequence. Thus, sequences over fields for additive stream ciphering and for some code-division multiple-access systems should have a large linear complexity.

For sequences over $\mathbf{Z}/(m)$, the Berlekamp-Massey algorithm does not work, but the Reeds-Sloane algorithm described in Section 4.4 works, and it is also efficient [70]. Thus, it is necessary to control the linear complexity of sequences over $\mathbf{Z}/(m)$ for additive synchronous stream ciphering whose basic idea is described as follows.

Both the sender and receiver have the same generator which produce a keystream sequence over an Abelian group $(G, +)$. Let $\mathbf{m} = m_1 m_2 \cdots m_t$ be a piece of plaintext, where $m_i \in G$. Then the sender uses the generator to produce a piece of keystream sequence $\mathbf{k} = k_1 k_2 \cdots k_t$. Then he or she encrypts the message $\mathbf{m}$ as

$$\begin{aligned} \mathbf{c} &= \mathbf{m} + \mathbf{k} = (m_1 + k_1)(m_2 + k_2) \cdots (m_t + k_t) \\ &= c_1 c_2 \cdots c_t, \end{aligned}$$

where $c_i = (m_i + k_i)$. The receiver uses the generator to produce the same piece of keystream $\mathbf{k}$, and then decrypts the message as $\mathbf{m} = \mathbf{c} - \mathbf{k}$. This is the basic idea of additive synchronous stream ciphering.

The study of sequences over $\mathbf{Z}/(m)$ has a long history [99]. So far some results about the period of linear recurrence sequences over $\mathbf{Z}/(m)$ are known, but little about the linear complexity of sequences over $\mathbf{Z}/(m)$ has been achieved. Sequences over fields are easy to construct, and their properties are easy to control. But it looks hard to do so for sequences over residue class rings. In this section we show how to construct sequences over $\mathbf{Z}/(m)$ from those over finite fields $\mathbf{Z}(p)$, where p is a prime, and how to control their cryptographic properties with the help of the CRT.

An important result we need is the following theorem whose second part was implied in the work of Reeds and Sloane [70] without giving proofs.

Theorem 7.3.1 *Let s^∞ be a sequence over $\mathbf{Z}/(m)$, where $m = m_1 m_2 \cdots m_t$ and m_i are pairwise relatively prime, and let*

$$s(i)^\infty = s^\infty \bmod m_i, \quad i = 1, 2, ..., t,$$

i.e., $s(i)_j = s_j \bmod m_i$ for all possible j.

1. *If s^∞ is (ultimately) periodic, then each sequence $s(i)$ must be (ultimately) periodic, and $\mathrm{per}(s^\infty) = \mathrm{lcm}\{\mathrm{per}(s(1)^\infty), \cdots, \mathrm{per}(s(t)^\infty)\}$, where $\mathrm{per}(s^\infty)$ denotes the least period.*

2. *$L(s^\infty) = \max\{L(s(1)^\infty), \cdots, L(s(t)^\infty)\}$, where $L(s^\infty)$ denotes the linear complexity.*

Proof: Let φ be the mapping from $\mathbf{Z}/(m)$ to $\mathbf{Z}/(m_1) \times \cdots \times \mathbf{Z}(m_t)$ given by

$$\varphi : x \bmod m \mapsto (x \bmod m_1, ..., x \bmod m_t).$$

By the CRT φ is an isomorphism.

Assume that s^∞ is periodic and $N = \mathrm{per}(s^\infty)$. Then $\varphi(x_{N+j}) = \varphi(x_j)$ for all $j \geq 0$. It follows that $s(i)_{N+j} = s(i)_j$ for all $j \geq 0$ and all $i = 1, ..., t$. Thus, each $s(i)^\infty$ is periodic and $N' = \mathrm{lcm}\{\mathrm{per}(s(1)^\infty), \cdots, \mathrm{per}(s(t)^\infty)\}$ divides N. On the other hand, since $s(i)_{N'+j} = s(i)_j$ for all j and i, we have $\varphi(s_{j+N'}) = \varphi(s_j)$ for all j. Since φ is one-to-one, $s_{j+N'} = s_j$ for all j. It follows that N divides N'. Combining the above results gives $N = N'$.

Now we prove part two. Let $l = L(s^\infty)$ and

$$s_j = a_1 s_{j-1} + a_2 s_{j-2} + \cdots + a_l s_{j-l}, \quad j \geq l \qquad (7.8)$$

be a shortest linear recurrence relation of s^∞, where $a_i \in \mathbf{Z}/(m)$. Let $\varphi(a_i) = (a(1)_i, \cdots, a(t)_i)$ for $i = 1, ..., t$. Applying the isomorphism φ to Equation (7.8) gives

$$\varphi(s_j) = \varphi(a_1)\varphi(s_{j-1}) + \cdots + \varphi(a_l)\varphi(s_{j-l}), \quad j \geq l$$

from which it follows that for each i with $1 \leq i \leq t$

$$s(i)_j = a(i)_1 s(i)_{j-1} + \cdots + a(i)_l s(i)_{j-l}, \quad j \geq l,$$

where $a(i)_j \in \mathbf{Z}/(m_i)$. Thus, $l_i = \mathrm{L}(s(i)^\infty) \leq l$.

On the other hand, let $l' = \max\{l_1, ..., l_t\}$. Assume that

$$s(i)_j = a(i)_1 s(i)_{j-1} + \cdots + a(i)_{l_i} s(i)_{j-l_i}, \quad j \geq l_i$$

is a shortest linear recurrence relation the sequence $s(i)^\infty$ satisfies. Define $a(i)_j = 0$ for all j with $l_{i+1} \leq j \leq l'$, where $1 \leq i \leq t$. Then

$$\begin{aligned}
(s(1)_j, \cdots s(t)_j) &= (a(1)_1, \cdots a(t)_1)(s(1)_{j-1}, \cdots s(t)_{j-1}) + \cdots + \\
&\quad + (a(1)_{l'}, \cdots a(t)_{l'})(s(1)_{j-l'}, \cdots s(t)_{j-l'})
\end{aligned} \quad (7.9)$$

holds for each $j \geq l'$. Let $a_j = \varphi^{-1}(a(1)_j, \cdots, a(t)_j) \in \mathbf{Z}/(m)$. Applying the inverse isomorphism φ^{-1} to Equation (7.9) gives

$$s_j = a_1 s_{j-1} + a_2 s_{j-2} + \cdots + a_{l'} s_{j-l'}, \quad j \geq l'.$$

Thus, $l \leq l'$. Hence $l = l'$. This proves part two. $\square$

Another important result we need is the following. Let s^∞ be a sequence of period N over $\mathbf{Z}/(p^k)$, and $s(p)^\infty = s^\infty \bmod p$. Assume that $\mathrm{L}(s^\infty) = l$ and

$$s_i = a_1 s_{i-1} + a_2 s_{i-2} + \cdots + a_l s_{i-l}, \quad i \geq l$$

is a shortest linear recurrence relation of s^∞, then

$$s(p)_i = a(p)_1 s(p)_{i-1} + a(p)_2 s(p)_{i-2} + \cdots + a(p)_l s(p)_{i-l}, \quad i \geq l,$$

where $a(p)_i = a_i \bmod p$, and $s(p)_i = s_i \bmod p$. It follows that

$$\mathrm{L}(s^\infty) \geq \mathrm{L}(s(p)^\infty). \tag{7.10}$$

This inequality provides a bridge for transferring bounds on the linear complexity of sequences over $\mathbf{Z}/(p)$ to those of sequences over $\mathbf{Z}/(p^k)$.

Theorem 7.3.1 and bounds (7.10) show that to control the least period and linear complexity of sequences over $\mathbf{Z}/(m)$, it is sufficient to do so for sequences over fields $\mathbf{Z}/(p_i)$. Thus, bounds on the linear complexity of sequences over fields are easily transferred to those of sequences over $\mathbf{Z}/(m)$, where $m = \prod p_i^{e_i}$.

An important issue in stream ciphering is the design of keystream generators which produce pseudo-random sequences having certain properties such as large period, large linear complexity, ideal pattern distributions. To design a sequence generator over $\mathbf{Z}/(p_1 p_2 \cdots p_t)$, a set of sequence generators SGi over $\mathbf{Z}/(p_i)$ can be combined by the CRA. Let $s(i)^\infty$ be the semi-infinite sequence produced by generator SGi over $\mathbf{Z}/(p_i)$, $m = p_1 p_2 \cdots p_t$, where p_i are pairwise distinct primes. Furthermore, let $P_i = m/p_i$, and u_i be the integer such that $u_i P_i = 1 \bmod p_i$.

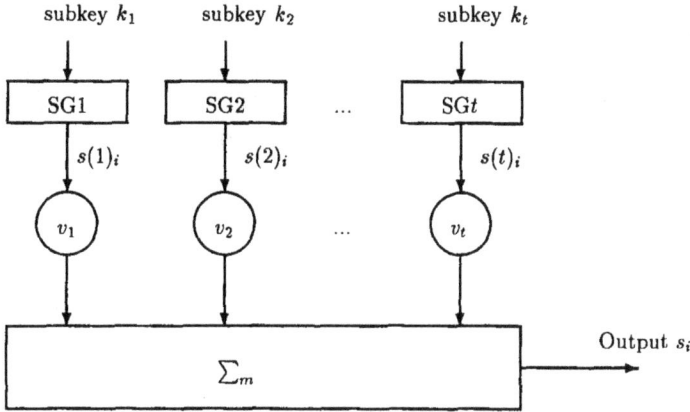

Figure 7.2: A sequence generator based on CRA.

Then a sequence generator over $\mathbf{Z}/(m)$ produces a semi-infinite sequence s^∞ with

$$s_j = \sum_{i=1}^{t} u_i P_i s(i)_j \bmod m, \quad i = 0, 1, \dots, \tag{7.11}$$

where $s(i)_j$ is the jth entry of the sequence $s(i)^\infty$. By controlling the properties of each field sequence $s(i)^\infty$, we can control those of the sequence s^∞. The sequence generator SG over $\mathbf{Z}/(m)$ is depicted by Figure 7.2 whose key consists of $k_1, k_2, \dots, k_t$, where k_i is the key of the generator SGi, $v_i = u_i P_i \bmod m$, and $\sum_m$ denotes integer addition modulo m.

7.4 CRA and Knapsack Problems

A *knapsack vector* $\mathbf{a} = (a_1, a_2, \dots, a_n)$ is an ordered n-tuple of distinct positive integers, where $n \geq 3$. The *knapsack problem* is to find a subset of the set $A = \{a_1, \dots, a_n\}$ whose elements sum up to a given integer. The instance of a knapsack problem is the knapsack problem with respect to a specific knapsack vector. For example, with respect to the specific knapsack vector

$$\mathbf{a} = (2, 4, 8, 16, 32, 64) \tag{7.12}$$

the knapsack problem is to find a vector $\mathbf{x} = (x_1, x_2, x_3, x_4, x_5, x_6)$ such that

$$\mathbf{a}\mathbf{x}^T = \sum_{i=1}^{6} a_i x_i = b \tag{7.13}$$

if the equation has solutions, where $x_i \in \{0,1\}$. If $b = 80$, this knapsack problem has only one solution $\mathbf{x} = (0,0,0,1,0,1)$.

The general knapsack problem is NP-complete, but some instances of the knapsack problem are easy to solve. This is the key idea behind the knapsack public-key cryptosystem [77, Chapter 3]. A knapsack vector $\mathbf{a} = (a_1, a_2, ..., a_n)$ is said to be *super-increasing* if

$$a_j > \sum_{i=1}^{j-1} a_i \qquad (7.14)$$

holds for all $j = 2, 3, ..., n$. The knapsack vector of (7.12) is super-increasing. The knapsack problem with respect to a super-increasing knapsack vector is easily solved as follows.

Let $\mathbf{a} = (a_1, a_2, ..., a_n)$ be super-increasing. Suppose that $x_1 a_1 + x_2 a_2 + \cdots + x_n a_n = b$. To solve the equation, we compare b with a_n. If $b > a_n$, then x_n must be 1; otherwise $x_n = 0$, since $a_n > \sum_{i=1}^{n-1} a_i$. This determines x_n with only one comparison between two integers. After this first step the problem is reduced to solving

$$x_1 a_1 + x_2 a_2 + \cdots + x_{n-1} a_{n-1} = b - x_n a_n.$$

Repeating the same procedure $n-1$ times gives the solution $\mathbf{x}$. Apparently, at most $n-1$ comparisons of integers and $n-1$ subtractions of integers are needed.

The Merkle-Hellman knapsack public-key cryptosystem is briefly described as follows [61]. A cryptosystem designer chooses a super-increasing knapsack vector $\mathbf{a} = (a_1, a_2, ..., a_n)$, a positive integer m such that

$$m > \sum_{i=1}^{n} a_i, \qquad (7.15)$$

and a positive integer t such that $\gcd(t,m) = 1$ and $t < m$. Then he or she computes $b_i = t a_i \bmod m$ for all i, and publicizes the vector $\mathbf{b} = (b_1, b_2, ..., b_n)$ as the encryption key. For a n-bit message block $\mathbf{m} = (m_1, m_2, ..., m_n)$ the corresponding ciphertext is

$$c = \sum_{i=1}^{n} m_i b_i.$$

The authorized receiver who knows t and m computes an integer u such that $ut = 1 \bmod m$ by Euclidean algorithm. After receiving c, he or she solves the knapsack problem

$$\sum_{i=1}^{n} m_i a_i = cu \bmod m,$$

which is the knapsack problem with respect to the super-increasing knapsack vector, and can be easily solved. A number of knapsack public-key crypto-systems turned out to be insecure, but they do serve as good examples to show the basic idea of public-key cryptography. For details about knapsack public-key cryptosystems, we refer to [77, Chapter 3]. Our aim of this section is to present other easy knapsack problems related to the CRA and CRT.

As we saw, the knapsack problem with respect to a super-increasing knapsack vector is easy to solve. But the knapsack vector of an easy knapsack problem is not necessarily super-increasing. For example, the vector

$$\mathbf{a} = (210, 330, 462, 770, 1155) \qquad (7.16)$$

is not super-increasing. But the knapsack problem with respect to this knapsack vector is very easy. To show this, let $m_1 = 11, m_2 = 7, m_3 = 5, m_4 = 3, m_5 = 2$. Then it is easy to check that $a_i \bmod m_j = 0$ for all $i \neq j$, and that $a_i \bmod m_i \neq 0$ for all i. Thus, if $\sum_{i=1}^{6} a_i x_i = b$, then it follows that $x_i = 0$ if $b \bmod m_i = 0$, and $x_i = 1$ otherwise. Thus, the knapsack problem with respect to this knapsack vector is easy.

We say that a knapsack vector $\mathbf{a} = (a_1, a_2, ..., a_n)$ is *orthogonal* if there are positive integers $m_1, m_2, ..., m_n$ such that

1. $a_i = 0 \bmod m_j$ for all pairs (i, j) with $i \neq j$;

2. $a_i \neq 0 \bmod m_i$ for all $i = 1, 2, ..., n$.

The vector of (7.16) is orthogonal. For the knapsack problem with respect to an orthogonal knapsack vector the following conclusion holds.

Theorem 7.4.1 *Let* $\mathbf{a} = (a_1, a_2, ..., a_n)$ *be an orthogonal knapsack vector. Then the knapsack problem with respect to* $\mathbf{a}$ *is solvable in linear time.*

Proof: By definition there are integers m_i such that $a_i \bmod m_j = 0$ for all $i \neq j$, and $a_i \bmod m_i \neq 0$ for all i. Assume that $\sum_{i=1}^{n} a_i x_i = b$, then $(a_i \bmod m_i)x_i = b \bmod m_i$. It follows that $x_i = 0$ if $b \bmod m_i = 0$, and $x_i = 1$ otherwise. Thus, exactly n modular operations gives the solution. □

It is easy to construct orthogonal knapsack vectors. Let $m_1, m_2, ..., m_n$ be pairwise relatively prime, where $m_i \geq 2$. Putting $m = \prod_{i=1}^{n} m_i$, and $a_i = m/m_i$ for all i, we then obtain an orthogonal vector $\mathbf{a} = (a_1, ..., a_n)$. The knapsack vector of (7.16) is actually constructed in this way. Now an interesting question is whether it is easy to find such a set of integers m_i, given an orthogonal vector $\mathbf{a} = (a_1, a_2, ..., a_n)$. The answer to this problem is "yes". Since $a_i = m/m_i$, it follows that

$$\prod_{i=1}^{n} a_i = m^{n-1}.$$

Thus, m is computed, and $m_i = m/a_i$ are known.

Theoretically, we can choose an orthogonal knapsack vector $\mathbf{a} = (a_1, a_2, ..., a_n)$, a positive integer M such that

$$M > \sum_{i=1}^{n} a_i,$$

and a positive integer $t < m$ such that $\gcd(t, M) = 1$. Then compute $b_i = ta_i \bmod M$, and publicize the vector $\mathbf{b} = (b_1, b_2, ..., b_n)$. In this way we have a knapsack public-key cryptosystem similar to the Merkle-Hellman system. But it looks that this system is insecure, just as the Merkle-Hellman system.

We now investigate the relation between the CRA and some knapsack problems. Recall that the Chinese remainder problem is to find the smallest non-negative integer u such that

$$u = r_i \bmod m_i, \quad i = 1, 2, ..., n,$$

where m_i are pairwise relatively prime. Let $m = m_1 m_2 \cdots m_n$, and $M_i = m/m_i$ as well as $u_i M_i = 1 \bmod m_i$. Then the CRA gives

$$u = \sum_{i=1}^{n} a_i x_i \bmod m, \qquad (7.17)$$

where $a_i = M_i u_i \bmod m$ for all i. The inverse of this problem is a modular knapsack problem, with knapsack vector $\mathbf{a} = (a_1, a_2, ..., a_n)$, if we allow x_i to be either 0 or 1 only. Furthermore, it is easily seen that $\mathbf{a}$ is orthogonal. This is in fact how the easy knapsack problem with respect to orthogonal knapsack vectors has been discovered.

It is interesting to note that the knapsack problem with respect to the super-increasing knapsack vector $\mathbf{a} = (1, 2, 2^2, ..., 2^{n-1})$ can be similarly solved. Assume that

$$\sum_{i=0}^{n-1} 2^i x_i = b.$$

Then $x_0 = b \bmod 2$, $x_1 = (b - x_0)/2 \bmod 2$, $x_2 = (b - x_0 - 2x_1)/4 \bmod 2$, and others can be similarly computed.

Even for some non-orthogonal knapsack vectors the knapsack problem may also be easy to solve. For example, let $\mathbf{a} = (a_1, a_2, ..., a_n)$ be a knapsack vector. It is natural to assume that $\gcd(a_1, a_2, ..., a_n) = 1$. If $\gcd(a_2, ..., a_n) = d \geq 2$, then $\gcd(d, a_1) = 1$. Assume that $\prod_{i=1}^{n} a_i x_i = b$, then the bit x_1 must be 1 if $b \bmod d \neq 0$, and 0 otherwise. This determines x_1. Thus, the problem is reduced to

$$\sum_{i=2}^{n} a_i x_i = b - a_1 x_1.$$

Dividing this equation by d gives another one

$$\sum_{i=2}^{n} a_i' x_i = b',$$

where $a' = a/d$, and $b' = (b - a_1 x_1)/d$. If we can find an integer i such that $\gcd(a_2', ..., a_{i-1}', a_{i+1}', ..., a_n') = d' \geq 2$, repeating the same procedure determines x_i. Generally, let $\gcd(a_{i_1}, a_{i_2}, ..., a_{i_s}) = d_1 > 1$ and $\gcd(a_{i_{s+1}}, a_{i_{s+2}}, ..., a_{i_n}) = d_2 > 1$. Then $\gcd(d_1, d_2) = 1$, and the knapsack equation

$$\sum_{i=1}^{n} a_i x_i = b$$

is converted into the following modular knapsack equations

$$\sum_{i=1}^{s} (a_{i_j} \bmod d_2) x_{i_j} = b \bmod d_2$$

and

$$\sum_{i=s+1}^{n} (a_{i_j} \bmod d_1) x_{i_j} = b \bmod d_1.$$

Solving these two equations could be easier than solving the original one. This could be generalized to give methods which work efficiently for a large number of knapsack vectors.

To make the knapsack problem difficult to solve by this method, it might be sufficient to choose a knapsack vector $\mathbf{a} = (a_1, a_2, ..., a_n)$ such that a_i are pairwise relatively prime.

7.5 Public-Key Systems via CRT

In the preceding section we have considered a knapsack problem based on the CRT. An insecure public-key cryptosystem was mentioned there. In this section we describe some other cryptosystems based on the CRT. Security questions are left to the reader.

A knapsack public-key cryptosystem based on the CRT was suggested in 1988 by He and Lu [41]. It is based on the modular CRT knapsack problem and is described as follows.

Choose randomly k integers $m_1, \cdots, m_k$ such that $\gcd(m_i, m_j) = 1$, where $1 \leq i < j \leq k$. First, compute $m = m_1 m_2 \cdots m_k$ and $M_i = m/m_i$ for $1 \leq i \leq k$. Second, with Euclidean algorithm compute integers M_i' such that $1 \leq M_i' < m_i$ and $M_i' M_i = 1 \bmod m_i$. Then let $a_i = M_i' M_i \bmod m$ for $1 \leq i \leq k$. Now choose randomly a positive integer w such that $\gcd(w, m) = 1$, and compute $a_i' = w a_i \bmod m$ for $1 \leq i \leq k$.

The plaintext block is $\mathbf{x} = (x_1, \cdots, x_k)$, where $0 \leq x_i < 10^l$ for $1 \leq i \leq k$, and where l is chosen such that $10^l \leq \min_{1 \leq i \leq k} m_i$.

Choose an integer $m' > (a_1' + a_2' + \cdots + a_k')(10^l - 1)$ and an integer w' such that $1 < w' < m'$ and $\gcd(w', m') = 1$. Then public parameters of the public-key cryptosystem are m' and

$$a_i'' = w' a_i' \bmod m', \ 1 \leq i \leq k.$$

The secret parameters of the system are $m_1, \cdots, m_k, m, w, w'$.

For each plaintext block $\mathbf{x} = (x_1, \cdots, x_k)$, the encryption is

$$\mathbf{y} = \sum_{i=1}^{k} a_i'' x_i \bmod m'.$$

Let $(w')^{-1}$ denote the inverse of w' modulo m'. To decrypt, compute first $y' := (w')^{-1} y \bmod m'$, then compute $y'' := w^{-1} y'$, where w^{-1} denotes the inverse of w modulo m. It is easily seen that

$$y'' = a_1 x_1 + \cdots + a_k x_k \bmod m.$$

Thus, $x_i = y'' \bmod m_i$ since $0 \leq x_i < m_i$.

This is the modular knapsack public-key cryptosystem, and it was broken in 1991 [106]. In 1991 and 1992 two other public-key cryptosystems based on the CRT were proposed. They are similar in nature.

The Cao-Li Public-Key System:

The public-key system proposed by Cao and Li is described as follows [19]. Choose n pairwise distinct primes $m_i = 3 \bmod 4$, and compute

$$m := \prod_{i=1}^{n} m_i, \ M_i := m/m_i, \ i = 1, ..., n.$$

With Euclidean algorithm compute integers M_i' such that $M_i'M_i = 1 \bmod m_i$, where $0 < M_i' < m_i$, $1 \leq i \leq n$. Let $\lambda_i = M_i'M_i$ for $1 \leq i \leq n$, and

$$\Lambda = \mathrm{diag}[\lambda_1, \lambda_2, ..., \lambda_n] = \begin{bmatrix} \lambda_1 & 0 & \cdots & 0 \\ 0 & \lambda_2 & \cdots & 0 \\ \vdots & \vdots & \vdots & \vdots \\ 0 & 0 & \cdots & \lambda_n \end{bmatrix}.$$

Choose $n \times n$ invertible matrices P and P_1 such that

1. the entries $p(ij)$ and $p_1(ij)$ of the two matrices are nonnegative integers;

2. $p(ij) = 0 = p_1(ij)$ if $i < j$;

3. each entry is no greater than

$$\beta = \min_{1 \leq i \leq n} \left[\sqrt{\frac{m}{i(i+1)d}} \right],$$

where $d \geq 1$ is a chosen positive integer.

Let $A = P^T \Lambda P$ and $B = P_1^T A P_1$. The public parameter of the system is the matrix B, and the secret ones are P, P_1, and m_i for $1 \leq i \leq n$.

The plaintext block is $\mathbf{x} = (x_1, \cdots, x_n)$, where $0 \leq x_i \leq d$. The encryption is simply computed as

$$y = \mathbf{x} B \mathbf{x}^T.$$

Note that

$$y = \mathbf{x} P_1^T A P_1 \mathbf{x}^T = \mathbf{x} P_1^T P^T \Lambda P P_1 \mathbf{x}^T.$$

Let $\mathbf{z} = \mathbf{x} P_1^T P^T$. Then $y = \mathbf{z}\mathbf{z}^T = \lambda_1 z_1^2 + \cdots + \lambda_n z_n^2$. Hence

$$z_i^2 = y \bmod m_i, \quad i = 1, ..., n. \tag{7.18}$$

For each i compute the solution z_i of (7.18) such that $0 \leq z_i < m_i/2$. The decryption is finally carried out as

$$\mathbf{x} = \mathbf{z}(P^{-1})^T (P_1^{-1})^T.$$

As regards the design of the system, it is noted that finding the matrices P, P_1, and the moduli m_i might be easy. But one has to coordinate the choices of these parameters in order that the above three conditions are satisfied. In the system, solving equation (7.18) is necessary, an efficient algorithm exists when each m_i is a prime of the form $p = 3 \bmod 4$. For the efficient algorithm we refer to [47, pp. 47-49].

One security aspect of this system depends on whether the following problem can be solved:

Problem: Given an integral matrix H, does there exist another integral invertible matrix Q such that the matrix

$$J := (Q^{-1})^T H Q^{-1} = \text{diag}[\gamma_1, \cdots, \gamma_n]$$

is diagonal and that $\gamma_i \bmod m_j = 0$ if $i \neq j$, and $\gamma_i \bmod m_j = 1$ otherwise, where γ_i are integers. If there is such a matrix Q, how can it be computed?

If an efficient algorithm for finding the matrix Q is found, the system is then broken. The following public-key system based on the CRT seems much better as far as the efficiency is concerned.

The Zheng Public-Key System:

A similar public-key cryptosystem was proposed by Zheng as follows [107]. Choose an integral matrix $B = (b_{ij})_{n \times n}$, an integral invertible matrix $P = (p_{ij})_{n \times n}$ with $p_{ij} > 0$, and an arbitrary integral matrix $C = (c_{ij})_{n \times n}$ with $c_{ij} > 0$. Also choose n pairwise relatively prime integers m_i such that

$$m_i > \left(\sum_{k=1}^{n} p_{ki} \right)^2, \quad i = 1, ..., n. \tag{7.19}$$

Let $m = m_1 \cdots m_n$. Choose a positive integer W such that

$$W \geq n + \sum_{i=1}^{n} \sum_{j=1}^{n} c_{ij} + 1.$$

Then let

$$M = mW - 1 > 0, \quad \lambda_i = \frac{Wm}{m_i} \bmod M, \quad 1 \leq i \leq n.$$

Finally, let

$$\begin{aligned}
\Lambda &= \text{diag}[\lambda_1, \lambda_2, ..., \lambda_n] \\
A &= P \Lambda P^T + MB + C = (a_{ij})_{n \times n}.
\end{aligned}$$

The public parameter of the system is the matrix A, and the secret ones are P, m_i, M, and W.

The plaintext block of the system is $\mathbf{x} = (x_1, \cdots, x_n)$, where $x_i \in \{0, 1\}$. To encrypt, we compute

$$y = \mathbf{x} A \mathbf{x}^T = \sum_{i,j} a_{ij} x_i x_j.$$

To decrypt, let $\mathbf{z} = (z_1, \cdots, z_n) = \mathbf{x}P$. Then

$$
\begin{aligned}
y &= \mathbf{x}A\mathbf{x}^T = (\mathbf{x}P)\Lambda(\mathbf{x}P)^T + M\mathbf{x}B\mathbf{x}^T + \mathbf{x}C\mathbf{x}^T \\
&= \lambda_1 z_1^2 + \cdots + \lambda_n z_n^2 + M\mathbf{x}B\mathbf{x}^T + \mathbf{x}C\mathbf{x}^T.
\end{aligned}
$$

It follows that

$$
\begin{aligned}
my &= \frac{mWmz_1^2}{m_1} + \cdots + \frac{mWmz_n^2}{m_n} + mM\mathbf{x}B\mathbf{x}^T + m\mathbf{x}C\mathbf{x}^T \bmod M \\
&= \frac{mz_1^2}{m_1} + \cdots + \frac{mz_n^2}{m_n} + m\mathbf{x}C\mathbf{x}^T \bmod M.
\end{aligned}
$$

Note that

$$
\begin{aligned}
0 &\le \frac{mz_1^2}{m_1} + \cdots + \frac{mz_n^2}{m_n} + m\mathbf{x}C\mathbf{x}^T \\
&= \sum_{j=1}^{n} \frac{m}{m_j} \left(\sum_{k=1}^{n} x_k p_{kj} \right)^2 + m \sum_{i=1}^{n} \sum_{j=1}^{n} c_{ij} x_i x_j \\
&\le \sum_{j=1}^{n} \frac{m}{m_j} \left(\sum_{k=1}^{n} p_{kj} \right)^2 + m \sum_{i=1}^{n} \sum_{j=1}^{n} c_{ij} \\
&\le \sum_{i=1}^{n} \frac{mm_i}{m_i} + m \sum_{i=1}^{n} \sum_{j=1}^{n} c_{ij} \\
&= m \left(n + \sum_{i=1}^{n} \sum_{j=1}^{n} c_{ij} \right) \\
&\le m(W-1) = mW - m < mW - 1 = M.
\end{aligned}
$$

we have

$$
my \bmod M = \sum_{i=1}^{n} \frac{mz_i^2}{m_i} + m\mathbf{x}C\mathbf{x}^T.
$$

Let $y' = my \bmod M$. Since m_i are pairwise relatively prime, $\gcd(m/m_i, m_i) = 1$. Let h_i be the inverse of m/m_i modulo m_i. Then

$$
\begin{aligned}
h_i y' &= \sum_{i=1}^{n} \frac{h_i mz_i^2}{m_i} + m_i^{-1} m\mathbf{x}C\mathbf{x}^T \\
&= z_i^2 \bmod m_i.
\end{aligned}
$$

Notice also that $0 \le z_i$ and

$$
z_i^2 = \left(\sum_{k=1}^{n} x_k p_{ki} \right)^2 \le \left(\sum_{k=1}^{n} p_{ki} \right)^2 \le m_i.
$$

Hence

$$z_i = \sqrt{h_i y' \bmod m_i}, \ i = 1, ..., n.$$

Finally, we get the plaintext **x** by

$$\mathbf{x} = \mathbf{z}P^{-1}.$$

Compared with the Cao-Li system, the choice of the design parameters of this system is easier. The choice of B is very easy. It is also easy to choose the moduli m_i which satisfy equation (7.19). Also the encryption and decryption are efficient.

So far we have only described the two public-key systems based on the CRT. The evaluation of the security of these two cryptosystems is left to the reader.

Appendix A

Tutorial in Information Theory

Since information is an important topic of this book, we give here an introduction to the basics of information theory.

A *probabilistic model* may be viewed as an experiment with an outcome from a set of possible alternatives, with a probability measure on the alternatives. The set of possible alternatives is called the *sample space*. In this book we are only interested in the case that the sample space is finite. Thus, all the sample spaces in this appendix are finite. For a sample space $\{a_1, a_2, ..., a_K\}$, let $P(a_k)$ denote the probability that the outcome of the experiment is a_k. An *ensemble* or *probability space* is the collection of the sample space and the probability distribution, denoted as

$$U = \left\{ \begin{array}{cccc} a_1 & a_2 & \cdots & a_K \\ P_U(a_1) & P_U(a_2) & \cdots & P_U(a_K) \end{array} \right\}.$$

Suppose that we have an unbiased coin with one side having a 1 and the other having a 0. If we flip the coin, the alternatives of the outcome will be 0 and 1, and the probabilities should be $P_U(1) = P_U(0) = 1/2$. Thus, the ensemble is

$$U = \left\{ \begin{array}{cc} 1 & 0 \\ 1/2 & 1/2 \end{array} \right\}.$$

If we do the same flipping two times independently, we have a two-outcome experiment. The ensemble is

$$U = \left\{ \begin{array}{cccc} 00 & 01 & 10 & 11 \\ \frac{1}{4} & \frac{1}{4} & \frac{1}{4} & \frac{1}{4} \end{array} \right\}.$$

In a two-outcome experiment we denote the outcomes by x and y, and suppose that x is a selection from the set of alternatives $X = \{a_1, a_2, ..., a_K\}$, called the sample space for x, and y from $Y = \{b_1, b_2, ..., b_J\}$, called the sample space for y. The set $\{(a_k, b_j) : 1 \leq k \leq K, 1 \leq j \leq J\}$ is called the *joint sample space* of x and y, and

$$XY = \left\{ \begin{array}{ccccccc} (a_1, b_1) & \cdots & (a_1, b_J) & \cdots & (a_K, b_1) & \cdots & (a_K, b_J) \\ P(a_1, b_1) & \cdots & P(a_1, b_J) & \cdots & P(a_K, b_1) & \cdots & P(a_K, b_J) \end{array} \right\}$$

is called a joint XY ensemble, where $P(a, b)$ denotes $P_{XY}(a, b)$.

Within an ensemble or a joint ensemble, an *event* is defined to be a subset of elements in the sample space. The probability of an event is the sum of the probabilities of the elements in the subset comprising that event. For example, $\{00, 11\}$ is an event with respect to the two-time coin-flipping model, and the probability of this event is $1/4 + 1/4 = 1/2$. In the joint XY ensemble, the event that x assumes a particular value a_k corresponds to the subset $\{(a_k, b_1), (a_k, b_2), ..., (a_k, b_J)\}$. Thus the probability of this event is

$$P_X(a_k) = \sum_{j=1}^{J} P_{XY}(a_k, b_j). \tag{A.1}$$

This is, for short, written

$$P(x) = \sum_y P(x, y),$$

where x and y act as both variables and outcomes.

Similarly, the probability of a given y outcome is

$$P(y) = \sum_x P(x, y).$$

For example, in the above two-time coin flipping experiment let (x, y) denote the outcome. Then it is easily seen that

$$P_X(x = 1) = P_{XY}(x = 1, y = 0) + P_{XY}(x = 1, y = 1) = \frac{1}{2}$$

$$P_X(x = 0) = P_{XY}(x = 0, y = 0) + P_{XY}(x = 0, y = 1) = \frac{1}{2}$$

$$P_Y(y = 1) = P_{XY}(x = 1, y = 1) + P_{XY}(x = 0, y = 1) = \frac{1}{2}$$

$$P_Y(y = 0) = P_{XY}(x = 1, y = 0) + P_{XY}(x = 0, y = 0) = \frac{1}{2}$$

To introduce mutual information, we need the concept of the conditional probability. If $P_X(a_k) > 0$, the *conditional probability* that outcome y is b_j, given that outcome x is a_k, is defined as

$$P_Y(b_j|a_k) = \frac{P_{XY}(a_k, b_j)}{P_X(a_k)}, \tag{A.2}$$

which is abbreviated as

$$P(y|x) = \frac{P(x,y)}{P(x)}.$$

Similarly,

$$P(x|y) = \frac{P(x,y)}{P(y)}.$$

The events $x = a_k$ and $y = b_j$ are said to be *statistically independent* if

$$P_{XY}(a_k, b_j) = P_X(a_k)P_Y(b_j), \tag{A.3}$$

which is equivalent to

$$P_{Y|X}(b_j|a_k) = P_Y(b_j)$$

if $P_X(a_k) > 0$. The ensembles X and Y are statistically independent if (A.3) is satisfied for all pairs (a_k, b_j) in the joint sample space.

Uncertainty, self-information and entropy

Let $\{a_1, ..., a_K\}$ be the sample space of an ensemble X with the probability assignment $P_X(a_k)$. If $P_X(a_k) = 1$ (and of course all other $P_X(a_i) = 0$), then the event $x = a_k$ gives us no information at all since we know that this event must happen. There is no uncertainty about the outcome of the experiment. Intuitively, the larger the probability $P_X(a_k)$, the less the uncertainty of the event $x = a_k$. Thus, it is natural to define the *self-information* and *uncertainty* of an event $x = a_k$ as

$$I(a_k) = -\log P_X(a_k) = \log \frac{1}{P_X(a_k)}, \tag{A.4}$$

where $P_X(a_k) > 0$. Consequently, we have

$$I(a_k) + I(a_j) = -\log P_X(a_k)P_X(a_j).$$

The base of the logarithm in the above definition determines the numerical scale used to measure information. For base 2 logarithms, the numerical value of (A.4) is called the number of *bits* (binary digits) of information, and for natural logarithms, it is called the number of *nats* (natural units) of information. These are the most common bases.

The *entropy* of an ensemble

$$U = \left\{ \begin{matrix} a_1 & a_2 & \cdots & a_K \\ P_U(a_1) & P_U(a_2) & \cdots & P_U(a_K) \end{matrix} \right\}$$

is defined to be the average uncertainty of each individual event $u = a_k$, i.e.,

$$\begin{aligned} H(U) &= \sum_{k=1}^{K} P_U(a_k) \log \frac{1}{P_U(a_k)} \\ &= -\sum_{k=1}^{K} P_U(a_k) \log P_U(a_k) \\ &= \sum_{k=1}^{K} P_U(a_k) I(a_k). \end{aligned}$$

Here and hereafter we take $0 \log 0$ to be 0. This corresponds to the limit of $h \log h$ as h approaches 0 from above. Thus, the entropy is also the average self-information of each individual event $u = a_k$. By definition entropy is a global property of an ensemble, while self-information is a local one. The entropy of an ensemble is closely related to the entropy used in statistical thermodynamics.

As an example, consider the following ensemble

$$U = \left\{ \begin{matrix} a_1 & a_2 & a_3 \\ \frac{1}{6} & \frac{2}{6} & \frac{3}{6} \end{matrix} \right\}.$$

By definition we have for the self-information $I(a_1) = \log 6$, $I(a_2) = \log 3$, $I(a_3) = \log 2$. The entropy is

$$H(U) = \frac{1}{6} \log 6 + \frac{2}{6} \log 3 + \frac{3}{6} \log 2 = \log 2^{\frac{2}{3}} 3^{\frac{1}{2}}.$$

To observe properties of the entropy function, we take the ensemble U with two outcomes. Let p and $1 - p$ denote the probability assignment of the two-outcome ensemble. Then the entropy function is

$$H_2(p) = -p \log p - (1 - p) \log(1 - p),$$

where $0 \leq p \leq 1$. By elementary calculus the function $H_2(p)$ is increasing in the interval $[0, 1/2]$, and decreasing in the interval $[1/2, 1]$. Apparently, $H_2(\frac{1}{2} + h) = H_2(\frac{1}{2} - h)$ for any h in $[0, 1/2]$. Thus, if $p = 1/2$, the entropy $H_2(1/2) = \log 2$, which is maximal.

Mutual information

Let $\{a_1, ..., a_K\}$ be the X sample space and $\{b_1, ..., b_J\}$ be the Y sample space in an XY joint ensemble with the probability assignment $P_{XY}(a_k, b_j)$. By the definition of uncertainty and self-information the *conditional uncertainty* and *conditional self-information* of the event $x = a_k$, given the occurrence of the event $y = b_j$, is

$$I(a_k|b_j) = -\log P_{X|Y}(a_k|b_j).$$

But the original uncertainty of the event $x = a_k$ is $-\log P_X(a_k)$. Thus, the information provided about the event $x = a_k$ by the occurrence of the event $y = b_j$ is the difference of the two uncertainties, i.e.,

$$
\begin{aligned}
I_{X:Y}(a_k; b_j) &= I(a_k) - I(a_k|b_j) \\
&= -\log P_X(a_k) + \log P_{X|Y}(a_k|b_j) \\
&= \log \frac{P_{X|Y}(a_k|b_j)}{P_X(a_k)}.
\end{aligned}
\tag{A.5}
$$

Interchanging the role of x and y, we find that the information provided about the event $y = b_j$ by the occurrence of $x = a_k$ is

$$I_{Y:X}(b_j; a_k) = \log \frac{P_{Y|X}(b_j|a_k)}{P_Y(b_j)}. \tag{A.6}$$

By the definition of conditional probabilities, the right-hand sides of (A.5) and (A.6) are identical, i.e.,

$$
\begin{aligned}
I_{Y:X}(b_j; a_k) &= \log \frac{P_{XY}(a_k, b_j)}{P_X(a_k) P_Y(b_j)} \\
&= \log \frac{P_{X|Y}(a_k|b_j)}{P_X(a_k)} \\
&= I_{X:Y}(a_k; b_j).
\end{aligned}
$$

Accordingly, $I_{X:Y}(a_k; b_j)$ is called the *mutual information* between the events $x = a_k$ and $y = b_j$.

To illustrate the above concepts, we now take an example. One basic component of communications systems is the channel, for example, a telephone line, a high frequency radio link, a space communications link, a computer network link, or a storage medium. One model of communications channels is the *binary symmetric channel* depicted by Figure A.1, which is memoryless. The input alphabet and output alphabet are $\{0, 1\}$. Due to noise the input bit is reproduced correctly at the channel output with some fixed probability $1 - \epsilon$ and is altered

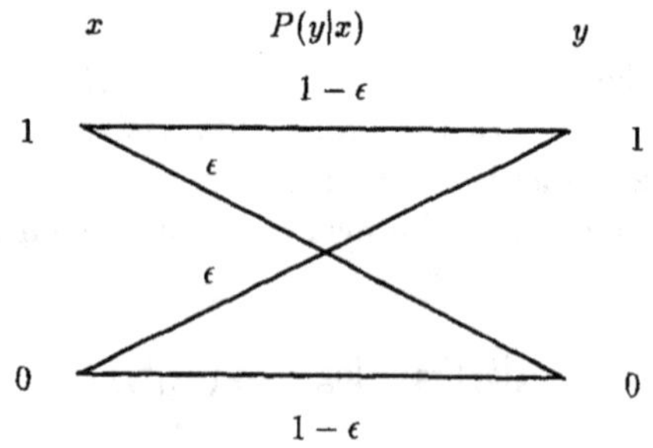

Figure A.1: The binary symmetric channel.

into the opposite bit with probability ϵ. Assume that $P_X(0) = P_X(1) = 1/2$. The joint probabilities are given by

$$P_{XY}(1,1) = P_{XY}(0,0) = \frac{1-\epsilon}{2},$$
$$P_{XY}(1,0) = P_{XY}(0,1) = \frac{\epsilon}{2}.$$

It is easy to see that $P_Y(0) = P_Y(1) = 1/2$. Thus, the self-information and uncertainty of the events $x = 0$, $x = 1$, $y = 0$ and $y = 1$ are $\log 2$.

By simple computation we get

$$P_{X|Y}(0|0) = P_{X|Y}(1|1) = 1 - \epsilon,$$
$$P_{X|Y}(1|0) = P_{X|Y}(0|1) = \epsilon.$$

By definition the mutual information is then

$$I_{X:Y}(0;0) = I_{X:Y}(1;1) = \log 2(1 - \epsilon),$$
$$I_{X:Y}(0;1) = I_{X:Y}(1;0) = \log 2\epsilon.$$

To have a better understanding of mutual information, we take another example. Suppose that the vector $\mathbf{x} = (x_1, x_2, ..., x_n)$ takes on each element of $GF(2)^n$ equally likely, where each variable x_i takes on 0 and 1 equally likely. Then the uncertainty and self-information of $\mathbf{x}$ is $\log_2 2^n = n$ bits. Let $\mathbf{v}_1$ be a nonzero vector of $GF(2)^n$. Then the event $\mathbf{v}_1\mathbf{x}^T = \sum_i v_{1,i}x_i = 0$ reduces the number of alternatives for $\mathbf{x}$ to 2^{n-1}. Thus, it provides $\log_2 2^n - \log_2 2^{n-1} = 1$ bit of information for $\mathbf{x}$. Let $\mathbf{v}_2$ be another vector of $GF(2)^n$. Then whether the event $\mathbf{v}_2\mathbf{x}^T = \sum_i v_{2,i}x_i = 0$ provides further information for $\mathbf{x}$ depends on whether the two vectors $\mathbf{v}_1$ and $\mathbf{v}_2$ are linearly independent. If they are linearly dependent, the second event provides no further information about $\mathbf{x}$; otherwise it provides one further bit information about $\mathbf{x}$. Generally, a set of simultaneous linear equations

$$\mathbf{v}_1\mathbf{x}^T = 0, \ \mathbf{v}_2\mathbf{x}^T = 0, \ \cdots, \ \mathbf{v}_m\mathbf{x}^T = 0 \tag{A.7}$$

provides altogether t bits of information about $\mathbf{x}$, where t is the rank of the matrix

$$V = \begin{bmatrix} \mathbf{v}_1 \\ \mathbf{v}_2 \\ \vdots \\ \mathbf{v}_m \end{bmatrix}.$$

As we have observed, the mutual information $I_{X|Y}(a_k, b_j)$ is only about the dependence between two events $x = a_k$ and $y = b_j$, and therefore is only a local property. The amount of average information one event in the the ensemble Y gives about an event in X ensemble is measured by

$$I(X;Y) = \sum_{k=1}^{K} \sum_{j=1}^{J} P_{XY}(a_k, b_j) \log \frac{P_{X|Y}(a_k|b_j)}{P_X(a_k)}.$$

This mean value is called the *average mutual information* with respect to X and Y.

Given the occurrence of $y = b_j$, the conditional mutual information $I_{X:Y}(x = a_k; y = b_j)$ may be different for different a_k. Thus, the average information the event $y = b_j$ gives to one event in the X ensemble is

$$
\begin{aligned}
I(X; b_j) &= \sum_{k=1}^{K} P_{X|Y}(a_k|b_j) I_{X:Y}(x = a_k; b_j) \\
&= -\sum_{k=1}^{K} P_{X|Y}(a_k|b_j) \frac{\log P_{X|Y}(a_k|b_j)}{P_X(a_k)}.
\end{aligned}
$$

This is called the average mutual information with respect to X and $y = b_j$. Similarly, we have the mutual information $I(a_k; Y)$ with respect to $x = a_k$ and Y.

If we take the average of $I(X; b_j)$ over the Y ensemble, we have

$$
\begin{aligned}
\sum_{j=1}^{J} P_Y(b_j) I(X; b_j) &= -\sum_{j=1}^{J} \sum_{k=1}^{K} P_Y(b_j) P_{X|Y}(a_k|b_j) \frac{\log P_{X|Y}(a_k|b_j)}{P_X(a_k)} \\
&= -\sum_{j=1}^{J} \sum_{k=1}^{K} P_{XY}(a_k|b_j) \frac{\log P_{X|Y}(a_k|b_j)}{P_X(a_k)} = I(X; Y).
\end{aligned}
$$

Similarly, we have

$$\sum_{k=1}^{K} P_X(a_k) I(a_k; Y) = I(X; Y).$$

So far we have several kinds of mutual information. They are averages of $I(a_k; b_J)$ over different ensembles.

In the example about binary symmetric channel above, the mutual information takes on the value $\log 2(1 - \epsilon)$ with probability $1 - \epsilon$ and the value $\log 2\epsilon$ with probability ϵ. The average mutual information is then given by $(1 - \epsilon) \log 2(1 - \epsilon) + \epsilon \log 2\epsilon$.

In most cases the amount of information only indicates how much we can get something done, but it says nothing about how to do it and also nothing about the complexity of the task. For instance, equation (A.7) only tells us the amount of information about the unknown $\mathbf{x}$, but it does not say how to find the solutions. Thus, in most cases information theory can direct us in solving problems, but it cannot give detailed techniques for problem solving.

Finally, we make an investigation of modular computation from the information point of view. Suppose that we have an unknown integer $\mathbf{x}$ in the range $[0, M - 1]$, where M is a positive integer. So far we have no other information about x, so the probability of x taking on each integer in the range is equally likely. Now suppose that we have the information that $x \bmod m_1 = b$, where and throughout the book $a \bmod m$ denotes the least nonnegative integer which is congruent to a modulo m. Observe, however, that we also use the standard number-theoretic congruence notation $a = b$ modulo m or $a \equiv b \pmod{m}$ to mean that m divides $a - b$. The meaning intended is clear from each context. How much information about x does the event $x \bmod m_1 = b$ provide?

If $m_1 \geq M$, then $x = b$. Thus, it provides $\log M$ information, which is the self-information and uncertainty of x. If $m_1 < M$, by the knowledge $x \bmod m_1 = b$ there must exist an integer $0 \leq k \leq \lfloor \frac{M-b}{m_1} \rfloor$ such that $x = km_1 + b$. Thus, the uncertainty of x reduces to $\log(\lfloor (M - b)/m_1 \rfloor + 1)$. Hence the information about x the event provides is

$$\log M - \log \left(\left\lfloor \frac{M - b}{m_1} \right\rfloor + 1 \right).$$

Further discussion about this aspect can be found in Section 3.6.

In this tutorial we have only considered the information aspect of a finite ensemble. This is the case we are concerned with in this book. For details about information theory we refer, for instance, to Gallager [34], and for algorithmic information we refer to [74].

Appendix B

Tutorial in Algebra

Notions and properties of groups, rings, ideals, quotient rings, Euclidean domains, fields, and other algebraic systems are frequently utilized in this book without giving a definition. The aim of this tutorial is to give a brief introduction to the algebraic notions and properties we need in the book, in order to make the book fully self-contained.

A mapping $f : S \to T$ is said to be *injective* or an *injection* or *one-to-one* if distinct elements of S have distinct images, i.e., $s \neq s'$ implies $f(s) \neq f(s')$. The mapping is called *surjective* or *onto* if every element of T is an image. A mapping is *bijective* or a *bijection* if it is both injective and surjective.

Let S be a set. A mapping $\mu : S^2 \to S$ is called a *binary operation* of S. We often use $x \cdot y$ to denote $\mu(x, y)$, and "$\cdot$" to denote a binary operation μ.

By a *monoid* we understand a set S with an element e and a binary operation "$\cdot$" such that

C1: $x \cdot (y \cdot z) = (x \cdot y) \cdot z$ for all $x, y, z \in S$;

C2: $x \cdot e = e \cdot x = x$ for all $x \in S$.

If $x \cdot y = y \cdot x$ for all $x, y \in S$, the monoid is said to be *commutative*. The element e is called the *unity element* or *neutral element*, and is denoted by 1 in many cases. Condition C1 is the *associative law*.

Two monoids S, T are said to be *isomorphic*, denoted by $S \cong T$, if there is a bijection $f : S \to T$, called an *isomorphism*, such that

1. $f(x \cdot y) = f(x) \cdot f(y)$ for all $x, y \in S$, where "$\cdot$" denotes the binary operation of both S and T;

2. $f(e_s) = e_t$, where e_s and e_t denote the unity element of S and T respectively.

A mapping $f : S \to T$, not necessarily bijective, but satisfying the above two conditions, is called a *homomorphism*, often briefly a *morphism*. If $S = T$, a homomorphism is called an *endomorphism*, and an isomorphism is called an *automorphism*.

The set $\mathbf{Z}$ of integers is a monoid under the integer multiplication with unity element 1, and under integer addition with unity element 0. The set $\mathbf{Z}^+$ of nonnegative integers is a monoid under both addition and multiplication. The set $\mathbf{Q}$ of real numbers is also a monoid under both addition and multiplication.

An element a of a monoid S is said to be *invertible* if there exists $b \in S$ such that

$$a \cdot b = b \cdot a = 1,$$

where 1 is the unity element of S. Such an element b must be unique, and is called the *inverse* of a and usually denoted as a^{-1}. The monoid $(\mathbf{Z}, \times)$ has only two invertible elements ± 1, but each element of the monoid $(\mathbf{Z}, +)$ is invertible.

A *group* is a monoid in which every element is invertible. If in a group $xy = yx$ for all x and y, it is said to be *commutative* or *Abelian*. If the number of elements of a group is finite, the group is said to be *finite*. A group G is *cyclic* if there is an element $a \in G$ such that each element of G can be written as a power of a.

The notions of homomorphism, isomorphism, endomorphism and automorphism defined for monoids carry over to groups as a special case. The monoid $(\mathbf{Z}, +)$ is a group, but $(\mathbf{Z}, \times)$ is not a group.

Let $(G, \cdot)$ be a group. A subset H of G is called a *subgroup* if it is a group under the same binary operation. Thus, H is a subgroup of G if and only if, for any $a, b \in H$, $a \cdot b \in H$ and $a^{-1} \in H$. The intersection of two subgroups of a group is still a subgroup. The group $(\mathbf{Z}, +)$ is a subgroup of $(\mathbf{Q}, +)$.

A *binary relation* on a set S is a set R of ordered pairs (x, y) of elements of S, that is, $x, y \in S$. Relations are often denoted by the symbol $\sim$. Thus, the fact that x and y are in the relation R is denoted by $x \sim y$. A relation $\sim$ on S is called an *equivalence relation* if the following conditions are satisfied:

E1: For every $x \in S$, $x \sim x$ (reflexive);

E2: For all $x, y \in S$, if $x \sim y$, then $y \sim x$ (symmetric);

E3: For all $x, y, z \in S$, if $x \sim y$ and $y \sim z$, then $x \sim z$ (transitive).

For the set $\mathbf{Z}$ of integers, we define a relation $\sim$ by

$$x \sim y \text{ iff } 2 | x - y.$$

This is obviously an equivalence relation. It partitions the set $\mathbf{Z}$ into two equivalence classes $\{z \in \mathbf{Z} : z \text{ even}\}$ and $\{z \in \mathbf{Z} : z \text{ odd}\}$. Generally, for an equivalence relation $\sim$ on S the *equivalence class* containing x is defined to be

$$S_x = \{y \in S : y \sim x\}.$$

Then two equivalence classes S_x and S_y are either disjoint or identical. Thus, S is the union of a number of equivalence classes. This forms a partition of S, i.e., S is the union of a number of disjoint equivalence classes S_x.

Let G be a group and H a subgroup of G. The subgroup H induces the following equivalence relations. The first one, denoted by $\sim_1$, is described by

$$x \sim_1 y \text{ if and only if } yx^{-1} \in H,$$

and the second, denoted by $\sim_2$, is described by

$$x \sim_2 y \text{ if and only if } x^{-1}y \in H.$$

It is easily checked that they are equivalence relations on G. Under $\sim_1$ the equivalence class containing x is the set $Hx = \{hx : h \in H\}$, which is called a *left coset*. Thus, the group G is partitioned into the union of left cosets:

$$G = \cup Hx.$$

Likewise, under $\sim_2$ the equivalence class containing x is xH, which is called a *right coset*. And the group is partitioned into the union of right cosets:

$$G = \cup xH.$$

A subgroup H of a group G is said to be *normal* if $Hc = cH$ for every $c \in G$, which is often written as $c^{-1}Hc = H$ for every $c \in G$. For a normal subgroup of G, let G/H denote the set of all right cosets. A binary operation for the set G/H is defined by

$$aH \cdot bH = abH.$$

It is easily verified that G/H is a group under this binary operation, which is called the *quotient group* of G by the normal subgroup H. We note that in an Abelian group every subgroup is normal. For any $n \in \mathbf{Z}$, $n\mathbf{Z}$ is a normal subgroup of $\mathbf{Z}$.

By a *ring* we understand a set R with two operations, $x + y$, called addition, and xy, called multiplication, such that

R1: R is an Abelian group under addition;

R2: R is a monoid under multiplication;

R3: addition and multiplication are related by the distributive laws:

$$(x + y)z = xz + yz, \quad z(x + y) = zx + zy.$$

The neutral element for addition is called *zero* and written 0, while the neutral element for multiplication is called *identity*, *unity* or *one* and written 1. R is called a *commutative ring* if $xy = yx$ for all $x, y \in R$. Two rings R_1 and R_2 are said to be *isomorphic* if there is a bijection $f : R_1 \to R_2$ such that

1. $f(x + y) = f(x) + f(y)$ for all $x, y \in R_1$;

2. $f(xy) = f(x)f(y)$ for all $x, y \in R_1$.

The mapping f is called an *isomorphism*. A *homomorphism* f is not necessarily bijective, and an *automorphism* satisfies $R_1 = R_2$.

An element a of a ring R is said to be *invertible* if it has an inverse in the multiplicative monoid R, i.e., if there exists an element $b \in R$ such that $ab = ba = 1$. A commutative ring in which every element other than 0 is invertible is called a *field*.

The set $\mathbf{Z}$ under the usual addition and multiplication is a commutative ring, but not a field, since only the two elements ± 1 are invertible. The set $\mathbf{Q}$ of rational numbers and the set $\mathbf{R}$ of real numbers are fields under the usual addition and multiplication.

An element $a \neq 0$ is called a *zero divisor* if there exists an element $b \neq 0$ such that $ab = 0$. A commutative ring without zero divisors is called an *integral domain*. The ring $(\mathbf{Z}, +, \times)$ is an integral domain. Every field is an integral domain.

Let R be an integral domain and suppose that with each $a \in R, a \neq 0$, a nonnegative integer $\mu(a)$, called the *norm*, is associated such that

U: for any $a, b \in R$, if $b \neq 0$, there exist $q, r \in R$ such that

$$a = bq + r,$$

where $\mu(r) < \mu(b)$ or $r = 0$.

Such an R is called an *Euclidean domain* or *Euclidean ring*. Let $\mu(x) = |x|$, the absolute value of an integer x, then $\mathbf{Z}$ is an Euclidean domain with respect to this norm.

For any field F, we consider formal sums $p(x) = p_0 + p_1 x + \cdots + p_k x^k$, where $p_i \in F$, called *polynomials* over F. Two polynomials are identical if and only if their corresponding coefficients are identical. Such a polynomial is in fact a *formal polynomial,* not an ordinary *polynomial function.* However, sums and products are defined as for ordinary polynomial functions. Let $F[x]$ denote the set of all polynomials over F. The addition of two polynomials

$$f(x) = \sum_{i=0}^{m} f_i x^i, \quad g(x) = \sum_{i=0}^{n} g_i x^i$$

is defined by

$$(f + g)(x) = \sum_{i=0}^{\max\{m,n\}} (f_i + g_i) x^i,$$

where f_i is defined to be zero if $i > m$, and g_i is defined to be zero if $i > n$. And the multiplication is defined by

$$(fg)(x) = \sum_{k=0}^{m+n} h_k x^k,$$

where $h_k = \sum_{i+j=k} f_i g_j$. With these two binary operations the set $F[x]$ is a ring. If we define $\mu(p)$ to be the degree of $p(x)$, $F[x]$ is an Euclidean domain with respect to this norm μ.

An *ideal* I of a ring R is a subgroup of the additive group of R such that $RI \subseteq I$ and $IR \subseteq I$, where $IR = \{ir : i \in I, r \in R\}$. Let f be a homomorphism from R_1 to R_2, the *kernel* of f is defined to be

$$\ker f = \{x \in R_1 : f(x) = 0\},$$

which is an ideal of R_1.

Consider a ring R and an ideal I in R, and denote by R/I the set of cosets of I in R. The set is again an Abelian group under the binary operation $(a + I) + (b + I) = (a + b) + I$. We can define a multiplication operation for R/I by

$$(a + I)(b + I) = ab + I.$$

Under the multiplication and the addition the set R/I becomes a ring, which is called the *quotient ring* of R by the ideal I. An ideal I of R is called *principal* if it is generated by a single element $r \in R$. An ideal generated by a single element r is written as (r). An integral domain is called a *principal ideal domain* if all of its ideals are principal. The rings $\mathbf{Z}$ and $F[x]$ are principal ideal domains. In fact, every Euclidean domain is principal.

Consider the ring $\mathbf{Z}$ under the usual addition and multiplication. For any positive integer m, the set $m\mathbf{Z}$ is an ideal of $\mathbf{Z}$, and is also written as (m). The quotient ring $\mathbf{Z}/m\mathbf{Z}$ consists of the elements $i + (m)$ for $i = 0, 1, ..., m - 1$. This ring is called the *residue class ring* of integers modulo m. Let $\mathbf{Z}_m = \{0, 1, ..., m - 1\}$. Define the addition and multiplication for $\mathbf{Z}_m$ as

$$x + y = (x + y) \bmod m, \quad xy = xy \bmod m,$$

where and throughout the book $a \bmod m$ denotes the least nonnegative integer which is congruent to a modulo m. Observe, however, that we also use the standard number-theoretic congruence notation $a = b$ modulo m or $a = b \ (\bmod \ m)$ to mean that m divides $a - b$. The meaning intended is clear from each context. Thus, $\mathbf{Z}_m$ forms a ring under the two binary operations. Defining the mapping $f : \mathbf{Z}/(m) \to \mathbf{Z}_m$ as

$$f(i + (m)) = i,$$

we know that this is an isomorphism. In this book we identify $\mathbf{Z}/(m)$ with $\mathbf{Z}_m$. From the theoretical point of view, $\mathbf{Z}/(m)$ could be more convenient, but $\mathbf{Z}_m$ is more convenient from the computational point of view. The ring $\mathbf{Z}/(m)$ is a field if and only if m is prime. If m is composite, the nontrivial factors of m give rise to zero divisors of $\mathbf{Z}/(m)$.

For a field F, if there is a positive integer p such that

$$0 = pa = a + a + \cdots + a, \text{ for all } a,$$

the least of all such p's is called the *characteristic* of F; otherwise its characteristic is defined to be zero. If the characteristic of a field is not zero, it must be a prime. The field $\mathbf{Q}$ of rational numbers has characteristic zero, while the residue class field $\mathbf{Z}/(p)$ has characteristic p, where p is a prime number.

A field F is said to be *finite* if the number of its elements is finite. The number of elements of a finite field must be p^m, where p is the characteristic of the field and k is a positive integer. Finite fields are called *Galois fields* and are usually denoted by $GF(p^m)$. Finite fields with p^m elements are isomorphic with each other.

The simplest finite fields are the residue class fields $\mathbf{Z}/(p)$, which are denoted by $GF(p)$. The field $GF(2) = \{0, 1\}$ satisfies: $0 + 1 = 1 + 0 = 1, 0 + 0 = 1 + 1 = 0, 0 \times 1 = 1 \times 0 = 0 \times 0 = 0, 1 \times 1 = 1$.

The field $GF(p^m)$ can be constructed as follows. Choose an irreducible polynomial $p(x)$ of degree m over $GF(p)$ and consider the ideal $(p(x))$ of $GF(p)[x]$. Since $p(x)$ is irreducible, it is not hard to prove that the residue class ring $GF(p)[x]/(p(x))$ is a field of p^m elements. In other words, the finite field $GF(p^m)$

can be described as follows. Let $GF(p^m)$ denote the set of all polynomials of degree less than m over $GF(p)$, i.e.,

$$GF(p^m) = \{a_0 + a_1 + \cdots + a_{m-1}x^{m-1} : a_i \in GF(p)\}.$$

The set $GF(p^m)$ forms a field under the addition and multiplication modulo the irreducible polynomial $p(x)$ of degree m.

For a finite field $GF(p^m)$ let $GF(p^m)^* = GF(p^m) \setminus \{0\}$. Then $GF(p^m)^*$ is a cyclic group of order $p^m - 1$, i.e., there is an element $\alpha \in GF(p^m)$ such that $GF(p^m)^* = \{1, \alpha, \alpha^2, \cdots, \alpha^{p^m-2}\}$. Such an element is called a *defining element* or *primitive element* of $GF(p^m)^*$. In the finite field $GF(p^m)$, for any elements x and y and any $k \geq 0$,

$$(x + y)^{p^k} = x^{p^k} + y^{p^k}.$$

Finally, we construct the finite field $GF(2^3)$. Observe that $\pi(x) = x^3 + x + 1 \in GF(2)[x]$ is irreducible over $GF(2)$, and consider α with $\pi(\alpha) = \alpha^3 + \alpha + 1 = 0$. Then the finite field $GF(2^3)$ consists of the following elements $\{0, 1, \alpha, \alpha^2, 1 + \alpha, \alpha + \alpha^2, 1 + \alpha + \alpha^2, 1 + \alpha^2\}$, where the operations are carried out modulo $\alpha^3 + \alpha + 1$.

Appendix C

List of Mathematical Symbols

$a \bmod m$ —The least nonnegative integer congruent to a modulo m or the residue class containing a modulo m. The meaning should be clear from the context.

$a(x) \bmod m(x)$ —The unique polynomial $b(x)$ congruent to $a(x)$ modulo $m(x)$ with $\deg(b(x)) < \deg(m(x))$ or the residue class with $a(x)$ modulo $m(x)$.

A^T —Transpose of the matrix A or the vector A.

$[a_0, ..., a_n]$—Continued fraction.

$\text{cont}(A)$ —Content of a polynomial.

$A \cong B$ —A is isomorphic to B.

$C^\perp$ —Dual code of a code C.

$\deg$ —Degree of polynomials.

$\text{diag}[a_1, \cdots, a_n]$—Diagonal matrix with a_i as the (i, i) entry.

$\gcd$ —Greatest common divisor.

$GF(q)$ —Galois field with q elements.

$GRR_k(M, \mathbf{e}, \mathbf{b})$ —Restricted generalized redundant residue code.

$GRS_k(\Delta, \mathbf{b})$ —Generalized Reed-Solomon code.

$H_2(p)$ —Entropy function of a probability space with two outcomes.

$H(U)$ —Entropy of probability space U.

$I(A)$ —Amount of self-information or uncertainty of event A.

$I(A; B)$ —Amount of mutual information between events A and B.

$\mathbf{L}(s)$ —Linear complexity or linear span of a sequence s.

lcm —Least common multiple.

ldcf —Leading coefficient of a polynomial.

$N(f)$ —Number of roots of a function f.

norm —Norm of an element.

$ord_n(q)$ —Multiplicative order of q modulo n.

$P(A)$ —Probability of event A.

$P(x|y)$ —Conditional probability.

pp(A) —Primitive part of polynomials.

res(A, B) —Resultant of two polynomials.

$R[x_1, ..., x_n\ ]$—Polynomial ring over R in n variables x_i.

$\mathbf{Z}/(m)$ —The residue class ring $\mathbf{Z}/(m)$ or the ring $\mathbf{Z}_m = \{0, 1, ..., m-1\}$ with usual addition and multiplication modulo m.

$\mathbf{Z}[x\ ]$—Polynomial ring over $\mathbf{Z}$.

$\mathbf{Z}$ —Ring of rational integers.

$\mathbf{Z}/(p)[x_1, ..., x_n\ ]$—Polynomial ring over $\mathbf{Z}/(p)$ in n variables x_i.

φ —Usually refers to the isomorphism in the CRT.

$\Phi(m)$ —Euler function.

$\mu(n)$ —Möbius function.

Bibliography

[1] R. C. Agarwal and C. S. Burrus, *Fast one-dimensional digital convolution by multidimensional techniques,* IEEE Trans. Acoustics, Speech, and Signal Processing, Vol. ASSP-**22**, No. **1** (1974), pp. 1-10.

[2] R. C. Agarwal and J. W. Cooley, *New algorithms for digital convolution,* IEEE Trans. Acoustics, Speech, and Signal Processing, Vol. ASSP-**25**, No. **5** (1977), pp. 392-410.

[3] R. Anderson, C. Ding, T. Helleseth and T. Kløve, *Democratic secret sharing with geometric codes,* to appear.

[4] Aryabhata, *Aryabhatiya,* K.S. Shukla's English Translation, 1979, Delhi, pp. 74-84.

[5] C. Asmuth and J. Bloom, *A modular approach to key safegarding,* IEEE Trans. Inform. Theory, Vol. IT-**29**, No. **2** (1983), pp. 208-210.

[6] F. Barsi and P. Maestrini, *Error detection and correction by product codes in residue number system,* IEEE Trans. Computers, Vol. C-**23**, No. **9** (1974), pp. 915-924.

[7] F. Barsi and P. Maestrini, *Improved decoding algorithm for arithmetic residue codes,* IEEE Trans. Inform. Theory, Vol. IT-**24**, No. **5** (1978), pp. 640-643.

[8] E. R. Berlekamp, *Algebraic Coding Theory,* New York: McGraw-Hill, 1968.

[9] C. Bingham, M. D. Godfrey, and J. W. Turkey, *Modern techniques of power spectral estimation,* IEEE Trans. Audio Electroacoust., Vol. AU-**15** (1967), pp. 56-66.

[10] R. E. Blahut, *Theory and Practice of Error-Control Codes,* Addison-Wesley Publishing Company, 1985.

[11] I. F. Blake, *Codes over certain rings*, Inform. Control **20** (1972), pp. 396-404.

[12] I. F. Blake, *Codes over integer residue rings*, Inform. Control **29** (1975), pp. 295-300.

[13] I. Borosh and A. S. Frankel, *Exact solutions of linear equations with rational coefficients by congruence techniques*, Math. Comp. **20**, 93 (Jan. 1966), pp. 107-112.

[14] D. C. Bossen and S. S. Yau, *Redundant residue polynomial codes*, Information and Control **13** (1968), pp. 597-618.

[15] W. S. Brown, *The completed Euclidean algorithm*, Bell Telephone Laboratories Rep., Murray Hill, N. J., June 1968.

[16] W. S. Brown and J. F. Traub, *On Euclid's algorithm and the theory of subresultants*, J. ACM, Vol. **18**, No. **4** (1971), pp. 505-514.

[17] W. S. Brown, *On Euclid's algorithm and the computation of polynomial greatest common divisors*, J. ACM, Vol. **18**, No. **4** (1971), pp. 478-504.

[18] D. A. Burgess and P. D. T. A. Elliot, *The average of the least primitive root*, Mathematika **15** (1968), pp. 39-50.

[19] Z. F. Cao and Y. C. Li, *A matrix-covering public-key cryptosystem* (in Chinese), Research Report of the Harbin University of Industry, 1991, pp. 1-37.

[20] L. Carlitz, *Distribution of primitive roots in a finite field*, Quart. J. Math. **4** (1953), 4-10.

[21] G. E. Collins, *Subresultants and reduced polynomial remainder sequences*, J. ACM **14** (1967), pp. 128-142.

[22] G. E. Collins, *The Calculation of Multivariate polynomial resultants*, J. ACM **18**, No. **4** (1971), pp. 515-532.

[23] J. W. Cooley and J. W. Tukey, *An algorithm for the machine calculation of complex Fourier series*, Math. Comput. **19** (1965), pp. 297-301.

[24] C. M. Cordes, *Permutations mod m in the form x^n*, Amer. Math. Monthly **83** (1976), no. **1**, pp. 32-33.

[25] Z. Dai, T. Beth, D. Gollmann, *Lower bounds for the linear complexity of sequences over residue rings*, Advances in Cryptology, Eurocrypt'90, Damgaard Ed., LNCS **473**, 1990, Springer, pp. 189-195.

[26] Z. Dai and K. C. Zeng, *Continued fractions and the Berlekamp-Massey algorithm*, Auscrypt '90, Springer Lecture Notes in Comp. Sci. vol. **453**, Springer Verlag, N. Y., 1990.

[27] B. Datta and A. N. Singh, *History of Hindu Mathematics*, 1938, Lahole, Vol. **2**, pp. 95-99.

[28] P. J. Davis and R. Hersh, *The Mathematical Experience*, Birkhäuser, Boston, 1981.

[29] L. E. Dickson, *History of the Theory of Numbers*, Vol. **1-3**, Chelsea Publishing Company, 1952.

[30] C. Ding, G. Xiao, W. Shan, *The Stability Theory of Stream Ciphers*, LNCS **561**, Springer-Verlag, 1991.

[31] C. Ding, *Binary Cyclotomic Generators*, Fast Software Encryption, LNCS **1008**, Springer-Verlag, 1995, pp. 29-61.

[32] P. D. T. A. Elliott, *The distribution of primitive roots*, Canad. J. Math. **21** (1969), 822-844.

[33] M. A. Fiol, *Congruences in Z^n, finite Abelian groups and the Chinese remainder theorem*, Discrete Math. **67** (1987), pp. 101-105.

[34] R. G. Gallager, *Information Theory and Reliable Communication*, New York: John Wiley and Sons, Inc., 1968.

[35] R. A. Games and A. H. Chan, *A fast algorithm for determining the complexity of a binary sequences with period 2^n*, IEEE Trans. Inform. Theory, IT-29 (1983), pp. 144-146.

[36] C. F. Gauss, *Disquisitiones Arith.*, Leipzig, 1801. English translation, Yale, New Haven, 1966. (Reprint by Springer-Verlag, Berlin, Heidelberg, and New York, 1986).

[37] S. W. Golomb, *Shift Register Sequences*, Aegean Park Press, Lguna Hills, California, 1982.

[38] I. J. Good, *The interaction algorithm and practical Fourier analysis*, J. Roy. Statist. Soc., Ser. B, Vol. **20** (1958), pp. 361-375; addendum, Vol. **22** (1960), pp. 372-375.

[39] I. J. Good, *The relationship between two fast Fourier transforms*, IEEE Trans. Computers, Vol. C-20 (1971), pp. 310-317.

[40] D. R. Heath-Brown, *Artin's conjecture for primitive roots*, Quart. J. Math. Oxford Ser. (2) **37** (1986), pp. 27-38.

[41] J. M. He and K. C. Lu, *The design and security of knapsack public-key cryptosystems* (in Chinese), Journal of Qinhua University, Vol. **28**, No. 1 (1988), pp. 89-97.

[42] L. Hua, *Introduction to Number Theory*, Berlin: Springer-Verlag, 1982.

[43] K. Ireland and M. Rosen, *A Classical Introduction to Modern Number Theory*, Second Edition, New York: Springer-Verlag.

[44] N. Jacobson, *Lectures in Abstract Algebra, Vol. III: Theory of Fields and Galois Theory*, Van Nostrand, Princeton, N. J., 1964.

[45] J. Justesen, *A class of constructive asymptotically good algebraic codes*, IEEE Trans. Inform. Theory **18** (1972), pp. 652-656.

[46] D. Knuth, *The art of computer programming, Vol. **2**. Seminumerical algorithms*, Addison-Wesley, Reading MA, 1981.

[47] N. Koblitz, *A Course in Number Theory and Cryptography*, Springer-Verlag, 1985.

[48] S. Y. Ku and R. J. Adler, *Algebra-based methods for solving multivariate polynomial equations*, Chemical Eng. Sci. Division Rep., No. **10-22-68**, Case Western Reserve U., 1968.

[49] S. Y. Ku and R. J. Adler, *Computing polynomial resultants: Bezout's determinant vs. Collins' reduced P.R.S. algorithm*, Comm. ACM **12**, 1 (Jan. 1969), pp. 23-30.

[50] R. Lidl, H. Niederreiter, *Finite Fields*, in Encyclopedia of Mathematics and Its Applications, vol. **20**, Addison-Wesley, 1983.

[51] F. J. MacWilliams, N. J. A. Sloane, *The Theory of Error-Correcting Codes*. North-Holland Publishing Company, 1977.

[52] D. M. Mandelbaum, *Some results in decoding of certain maximal-distance and BCH codes*, Information and Control **20** (1972), pp. 232-243.

[53] D. M. Mandelbaum, *Error correction in residue arithmetic*, IEEE Trans. Computers, Vol. C-**21**, No. **6** (1972), pp. 538-545.

[54] D. M. Mandelbaum, *On a class of arithmetic codes and a decoding algorithm*, IEEE Trans. Inform. Theory (Jan. 1976), pp. 85-88.

[55] D. M. Mandelbaum, *A method for decoding of generalized Goppa codes*, IEEE Trans. Inform. Theory, IT-**23** (1977), pp. 137-140.

[56] D. M. Mandelbaum, *Further results on decoding arithmetic residue codes*, IEEE Trans. Inform. Theory, Vol. IT-**24**, No. **5** (1978), pp. 643-644.

[57] J. L. Massey, *Shift-register synthesis and BCH decoding*, IEEE Trans. Inform. Theory, vol. IT-**15** (January, 1969), pp. 122-127.

[58] J. L. Massey, *Some applications of coding theory in cryptography*, in Codes and Cyphers: Cryptography and Coding IV (Ed. P. G. Farrell). Esses, England: Formara Ltd., 1995, pp. 33-47.

[59] M. T. McClellan, *The exact solution of systems of linear equations with polynomial coefficients*, J. ACM **20**, No. **4** (Oct. 1973). pp. 563-588.

[60] R. J. McEliece, *The Theory of Information and Coding*, Addison-Wesley, Reading, MA, 1977.

[61] R. Merkle and M. Hellman, *Hiding information and signatures in trapdoor knapsacks*, IEEE Trans. Inform. Theory, IT-**24** (1978), pp. 525-530.

[62] W. H. Mills, *Continued fractions and linear recurrence*, Math. Comp. vol. **29** (1975), pp. 173-180

[63] J. Moses, *Solution of systems of polynomial equations by elimination*, Comm. ACM **9**, 8 (1966), pp. 634-637.

[64] L. Murata, *On the Magnitude of the Least prime primitive Root*, J. Number Theory **37** (1991), pp. 47-66.

[65] O. Ore, *The general Chinese remainder theorem*, Amer. Math. Monthly **59** (1952), pp. 365-370.

[66] N. J. Patterson, *The algebraic decoding of Goppa codes*, IEEE Trans. Inform. Theory, IT-**21** (1975), pp. 203-207.

[67] W. Nöbauer, *Über Permutationspolynome und Permutations-funktionen für Primzahlpotenzen*, Monatsch. Math. **69** (1965), pp. 230-238.

[68] C. M. Rader, *Discrete convolution via Mersenne transforms*, IEEE Trans. Comput., Vol. C-**21** (1972), pp. 1269-1273.

[69] A. Ralston, *A First Course in Numerical Analysis*, McGraw-Hill, New York, 1965.

[70] J. A. Reeds and N. J. A. Sloane, *Shift-register synthesis (modulo m)*, SIAM J. Comput. **14**, No. 3 (1985), pp. 505-513.

[71] R. L. Rivest, A. Shamir, L. Adleman, *A method for obtaining signature and public-key cryptosystems*, Comms. of ACM **21**, No. 2, 1978, pp. 120-126.

[72] R. L. Rivest, *Remarks on a proposed cryptanalytic attack on the M.I.T. public-key cryptosystem*, Cryptologia **2**, 1978.

[73] J. B. Rosser, *A method of computing exact inverses of matrices with integer coefficients*, J. Res. Nat. Bureau Standards, Vol. **49** (1952), pp. 349-358. MR **14**, 1128.

[74] G. Rozenberg and A. Salomaa, *Cornerstones of Undecidability*, Prentice Hall, 1994, Chapter 5.

[75] A. Salomaa, *On essential variables of functions, especially in the algebra of logic*, Ann. Acad. Scient. Fennicae, Ser. AI 339 (1963).

[76] A. Salomaa, *Computation and Automata*, New York, Cambridge: Cambridge University Press, 1985.

[77] A. Salomaa, *Public-key Cryptography*, EATCS Monographs on Theoretical Computer Science, vol. **23**, Springer-Verlag, 1990.

[78] D. V. Sarwate, *On the complexity of decoding Goppa codes*, IEEE Trans. Inform. Theory, IT-**23** (1977), pp. 515-516.

[79] W. M. Schmidt, *Equations over Finite Fields: An Elementary Approach*, Lecture Notes in Mathematics **536**, Berlin: Springer-Verlag.

[80] A. Schönhage, *Multiplikation großer Zahlen*, Computing **1** (1966), pp. 182-196.

[81] A. Shamir, *How to share a secret*, Comm. ACM **22** (1979), pp. 612-613,

[82] P. Shankar, *On BCH codes over arbitrary integer rings*, IEEE Trans. Inform. Theory, IT-**25** (1979), pp. 480-483.

[83] K. Shen, *Historical development of the Chinese remainder theorem*, Archive for History of Exact Science **38** (1988), pp. 285-305.

[84] V. Shoup, *Searching for primitive roots in finite fields*, Math. Comput. **58**, no. 197 (1992), pp. 369-380.

[85] T. Siegenthaler, *Correlation-immunity of nonlinear combining functions for cryptographic application*, IEEE Trans. Inform. Theory, Vol. IT-**30**, No. 5 (Sept. 1984), pp. 776-780.

[86] N. J. A. Sloane, *Encryption by random rotations*, Proc. Workshop on Cryptography, LNCS **149**, Springer-Verlag, New York, 1983, pp. 71-128.

[87] E. Spiegel, *Codes over Z_m*, Inform. Control, **35** (1977), pp. 48-51.

[88] E. Spiegel, *Codes over Z_m, revisited*, Inform. Control **35** (1977), pp. 48-51.

[89] C. N. Srihivasiengar, *The History of Ancient India Mathematics*, 1967, Calcutta, pp. 101-102.

[90] J. Stone, *Multiple-burst error correction with the Chinese remainder theorem*, J. Soc. Indust. Appl. Math. **11** (1963), pp. 74-81.

[91] Y. Sugiyama, M. Kasahara, S. Hirasawa and T. Namekawa, *A method for solving key equation for decoding Goppa codes*, Inform. Control **27** (1975), pp. 87-99.

[92] M. Szalay, *On the Distribution of the primitive Roots of a prime*, J. Number Theory **7** (1975), pp. 184-188.

[93] L. H. Thomas, *Using a computer to solve problems in physics*, in "Applications of Digital Computers", Ginn and Co., Boston, MA, 1963.

[94] E. Vegh, *Pairs of Consecutive Roots Modulo a prime*, Proc. Amer. Math. Soc. **19**, 2, 1968, pp. 1169-1170.

[95] E. Vegh, *Arithmetic progressions of primitive roots of a prime. III*, J. reine angew. Math. **244** (1972), pp. 108-111.

[96] E. Vegh, *A Note on the Distribution of the Primitive Roots of a Prime*, T. Number Theory **3** (1971), pp. 13-18.

[97] B. L. Van der Waerden, *Modern Algebra*, Vol. **1**, Ungar, New York, 1948.

[98] Y. Wang, *On the least primitive root of a prime*, Sci. Sinica **10** (1961), pp. 1-14.

[99] M. Ward, *The arithmetical theory of linear recurring sequences*, Transactions of the American Mathematical Society, Vol. **35** (July 1933), pp. 600-628.

[100] A. Weil, *Solutions of equations in finite fields*, Bull. Amer. Math. Soc. **55** (1949), 497-508.

[101] A. L. Whiteman, *A Family of Difference Sets*, Illinois J. Math. **6** (1962), pp. 107-121.

[102] L. R. Welch and R. A. Scholtz, *Continued fractions and Berlekamp's algorithm*, IEEE Trans. Info. Theory vol. **25** (1979), pp. 19-27.

[103] S. Winograd, *On computing the discrete Fourier transform*, Math. Comput. **32**, No. 141 (1978), pp. 175-199.

[104] G. Z. Xiao and J. L. Massey, *S spectral characterization of correlation-immune functions*, IEEE Trans. IT-**34** (May 1988), pp. 569-571.

[105] S. Yu, Q. Zhuang and K. Salomaa, *The state complexities of some basic operations on regular languages*, Theoretical Computer Science **125** (1994), pp. 315-328.

[106] Z. Z. Zhang, *Breaking a new knapsack public-key cryptosystem* (in Chinese), System Science and Mathematics, Vol. **11**, 1 (1991), pp. 91-96.

[107] B. D. Zheng, *A public-key cryptosystem of the form of hiding linear transforms* (in Chinese), Journal of Electronics, Vol. **29**, No. 4 (1992), pp. 21-24.

Index

www.ingramcontent.com/pod-product-compliance
Lightning Source LLC
Chambersburg PA
CBHW050639190326
41458CB00008B/2345